Gigahertz and Terahertz Technologies for Broadband Communications

For a listing of recent titles in the *Artech House Space Technology and Applications Library,* turn to the back of this book.

Gigahertz and Terahertz Technologies for Broadband Communications

Terry Edwards

Artech House
Boston • London

Library of Congress Cataloging-in-Publication Data
Edwards, T.C. (Terence Charles)
    Gigahertz and terahertz technologies for broadband communications / Terry Edwards.
      p.   cm.—(Artech House space technology and applications library)
    Includes bibliographical references and index.
    ISBN 1-58053-068-0 (alk. paper)
      1. Broadband communication systems.   I. Series

    TK5103.4.E33   2000
    621.382—dc21   99-052315
                                            CIP

British Library Cataloguing in Publication Data
Edwards, Terry
    Gigahertz and terahertz technologies for broadband communications.
    —(Artech House space technology and applications library)
    1. Broadband communication systems   I. Title
    621.3'82
    ISBN 1-58053-068-0

Cover design by Igor Valdman

© 2000 Artech House, Inc.
685 Canton Street
Norwood, MA 02062

International Standard Book Number: 1-58053-068-0
Library of Congress Catalog Card Number: 99-052315

10 9 8 7 6 5 4 3 2 1

To the ever-closest person to the author, who has the greatest understanding and absolutely unbelievable patience, his wife Patricia Adene.

Contents

Preface

T HIS BOOK has been written to provide within one cover a critical
appraisal of mainly broadband communications systems, both fixed
and mobile, in terms of those already existing and many planned systems for twenty-first century high-speed Internet services. Almost all of
these types of systems use gigahertz radio or light (terahertz) frequencies.
A major aim is the "demystification" of this sector of technology. In terms
of style the book follows the genre established by Ira Brodsky (*Wireless:
The Revolution in Personal Telecommunications*, Artech House, 1995), and
to some extent this is a follow-on book to John G. Nellist and Elliott M.
Gilbert (*Understanding Modern Telecommunications and the Information Superhighway*, Artech House, 1999).

Extensive use is made of photographs of systems, diagrams showing
various techniques, and graphs showing market developments and
important systems characteristics. Many large corporate players located
worldwide have provided excellent photographs of various installations
and modules.

If you are a chairman or a CEO of almost any corporation currently active in these or related sectors, or if you are planning to enter these high-growth fields, then this book is definitely for you. In addition, academics and those simply interested in advanced communications technology will find this material stimulating.

The approach is nonmathematical and not deeply technical. It is, however, critical, comparative, and, in parts, doubtless controversial.

Acknowledgments

MANY INDIVIDUALS AND CORPORATIONS around the world contributed vitally to the preparation of this book, and the opportunity is taken here to acknowledge them as far as possible.

The reviewer, commissioned by the publisher at an early point during the preparation of this book, proved to be a wise choice and his guidance and comments have been of great value. My thanks go out to him. Also, I should like to thank both Mike Webb and Dr. Julie Lancashire at Artech House (London offices) for their helpfulness and patience.

Several people provided information that formed sound background material. Among these are Dave Cowan of NERA (U.K.), who kindly supplied information on trellis-coded modulation, and Professor Peter Cochrane of BT Labs (U.K.), who also kindly supplied some information that formed a basis for some details in the book. Simon West of Barrington Lloyd International was also extremely helpful on many occasions. Bruce R. Elbert, Elliott M. Gilbert, and John G. Nellist are also thanked because several charts and figures from their books are reused here.

Corporations who kindly contributed illustrations include:

- Barrington Lloyd International;
- City of Sunderland (United Kingdom);
- Federal City Communications Corporation and Angel Technologies Corporation;
- Fujitsu;
- GEC-Marconi;
- Hewlett-Packard;
- Hughes Electronics Corporation;
- Kingston TLI;
- Marconi Materials Technology;
- Nokia Mobile Phones;
- Nortel (Broadband Wireless Access);
- P-COM Inc.;
- Philips Electronics;
- Pirelli Cavi e Sistemi;
- Scientific-Atlanta;
- SkyBridge;
- ST Teleport Singapore;
- Teledesic LLC.

In the absence of any one of the above contributors, this book would have suffered considerably. Every one of those who contributed to this endeavor enhanced it greatly.

1

A Communications Revolution

1.1 Leaving basic POTS behind

In almost every respect the telecommunications scene repeatedly experiences revolutionary change. Many people are too young to remember Strowger exchanges, regularly crossed lines, and calls often failing to connect. Yet this was the case as recently as the 1980s even in many developed countries. For anyone living in an early-stage developing economy the introduction of the basic telephone and plain old telephone service (or POTS) is, of course, a major event which promises to transform lives and businesses.

The introduction of digital electronic exchanges, fiber-optic transmission, and satellite communications changed the scene radically. By the late twentieth century crossed lines had become a rare event in advanced economies and all callers expected their connections to be made the first time and to stay connected until hanging up—even at three cents per minute. Mobile phone use had grown almost exponentially and some people sported "dummy phones" so they would look really streetwise in spite of the fact they could not afford to use the real thing!

Total global telecommunications industry revenues, services, and equipment broke through the trillion dollar level in 1998 and (even allowing for the modest amount of Y2K disruption in year 2000) will approach $3.2 trillion by 2010. The overall trend is shown in Figure 1.1. Service revenues are always well ahead of equipment sales—witness the fact that mobile phones and PCs are often offered free of charge because the payback from the service side dominates so greatly.

As the world progresses through the early years of the third millennium so the fabric of national and global telecommunications becomes transformed. This transformation enables broadband fixed and mobile facilities to become a reality for rapidly increasing numbers of subscribers everywhere. In the United States alone broadband revenues are expected to reach $50 billion by 2005. Intelligent lightwave networks using dense-wavelength division multiplexing (DWDM), broadband satellite constellations, and seamlessly interconnecting wireless systems will all be linked into the self-healing and automatically rerouting global information superhighway—or "the supernet." Such networks are forming the principal on-ramps for this superhighway.

The original Internet—the darling of late twentieth-century communications—threatened the earlier preeminence of speech telephony traffic. The great popularity of e-mail and Web surfing meant that data traffic began to outstrip talking—certainly in the developed world. From

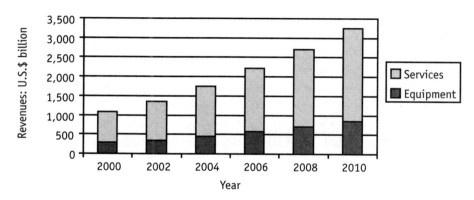

Figure 1.1 Global telecommunications revenues.

the old slogan, "it's good to talk," the transition was leading more toward, "it's good to net."

The growth of the Internet backbone market, on a global basis, is indicated in Figure 1.2. The basic technologies supporting this gigahertz and terahertz backbone include fiber optics, terrestrial millimeter-wave, and satellite communications systems.

People are, however, for the most part naturally social at all levels and this extends throughout personal, business, and professional life. This fact is likely to ensure that telephony has a highly significant future alongside the ever-expanding data needs.

Many industry analysts consider that the trend towards ever-increasing mobility will last "forever." According to data provided by the *Financial Times* (FT) of London, England, in 1997 mobile communications still only accounted for 19% of the total telecommunications market, the bulk (60%) being taken up by telephone services. The same source gave forecasts for the years 2001 and 2005, predicting that by 2005 34% of total revenues would be accredited to mobile. With some medical concerns surrounding mobile phone use, modest Y2K disruption, and likely stabilization instead of continuous growth in vehicular transportation, this may now be considered rather unlikely, and it is more probable that mobile use will only be proportionately somewhat higher than in 1997. On the other hand, advanced fixed services including broadband applications are likely to considerably increase their share. The substantially adjusted 2005 forecasts are shown in Figure 1.3.

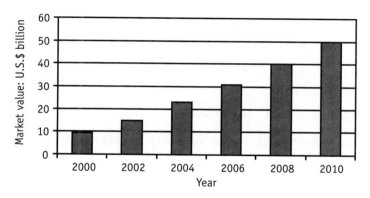

Figure 1.2 Global Internet backbone market.

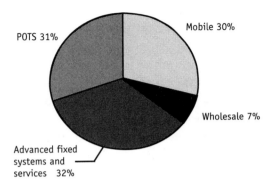

Figure 1.3 Telecommunications revenue breakdown for the year 2005.

In common with most advances in technology, the communications revolution all began with scientific and mathematical discoveries (see Table 1.1). We can usefully trace the developments back to Michael Faraday and James Clerk Maxwell of the nineteenth century.

In the "time travels" throughout the nineteenth and twentieth centuries presented in the table, the key people who match the key technological events are identified.

Well before the close of the twentieth century, true and complete convergence had been achieved between the computer and the telecommunications industries. Computer management and control were already ubiquitous in the following systems:

- Telephone services;

- Cable television (CATV—in most of the developed countries often including telephony and high-speed Internet access);

- Satellite [satellite TV—i.e., direct broadcast by satellite (DBS) and trunk satellite telecoms];

- High-speed fiber-optic connections—local and global;

- Broadband terrestrial wireless (LMDS, MVDS).

LMDS and MVDS stand for local multipoint distribution services and microwave video distribution services, respectively. These are described in Chapter 8.

Table 1.1

Key People and Key Technological Events

Some key people:		
1800s	**1900s**	**2000s**
Faraday, Maxwell, Morse	Alexander Graham Bell, Rainey (1926), Reeves (1939), Marconi, Arthur C. Clarke, Shockley et al. (1947), Kilby, Noyce (1958), Kao and Hockham (1966), Tim Berners-Lee (late 1980s)	Mainly teams rather than individual inventors
Some key events:		
1800s	**1900s**	**2000s**
Electricity Electromagnetic wave theory Transatlantic Telegraph	Telephone Pulse code modulation Practical radio Geostationary satellite Transistor Integrated circuit Fiber-optics theory First PC (Xerox) Mobile cellular communications The Internet	Mobile satellite Broadband satellite systems Broadband millimeter-wave (MVDS) Advanced semiconductors Photonic integrated systems Broadband mobiles Global information superhighway

The overall concept of this computer/telecommunications convergence is illustrated in Figure 1.4. For most of the time following Alexander Graham Bell's invention of the telephone, telecommunications continued to proceed using *analog* techniques. Signals were amplitude modulated, that is, "strength" modulated, onto carriers or "bearers" that extended in frequency up to the radio bands. Channels were separated by frequency-division multiplexing (FDM) and telephone messages were routed using electromechanical (Strowger) switch exchanges.

Radio transmission and reception proceeded using similar principles—although, later on, frequency-modulation (FM) gave greatly

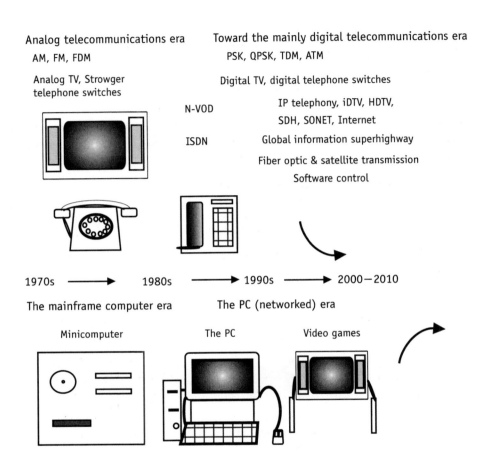

Figure 1.4 The convergence trend between telecommunications and computer systems.

improved radio reception quality. TV has remained steadfastly analog until comparatively recent times. Usable computers, of course, came along much later than the telephone, radio, or TV and no one could ever remotely consider owning a computer before the late 1970s (nobody had a room large enough, a pocket deep enough, or for that matter a need intense enough!). Through the 1970s minicomputers became vogue and interactive systems came into commercial use. Then, in the late 1970s, as we are all aware, nothing short of a revolution occurred with the commercial introduction of the PC.

Companies like Acorn, Commodore, IBM (the first to use the term *PC*), and many other contenders suddenly became household names and virtually everyone was talking computers. By the mid-1980s most school kids had a computer at their places of learning and mom and dad were becoming increasingly stressed-out with their children's ever more frequent requests for a PC at home. Games could be run on these amazing machines and sometimes even homework might be completed with the aid of that magical box of tricks!

The corporate and academic scenes became transformed by the new influx of PCs and soon these were becoming networked—connected together with a server that could manage such networks. Some of the PCs were little more than dumb terminals but this left the users feeling out of the mainstream so to speak. It's very interesting how most humans feel the personal need to remain in control of things and to have their own files of information.

By the late 1980s the world had transitioned from the "analog era" toward the mainly "digital telecommunications era." Almost all telephony switches (within the exchanges) were now electronic and digital, and the Integrated Services Digital Network (ISDN), conceived during the predominantly analog 1970s, was now enjoying a more realistic revival. Fiber-optic and satellite transmission technologies and their implementations continued a strong expansionary phase and this continues today.

Then, in the late 1980s, along came Tim Berners-Lee—the "father of the Web"—and the world of telecommunications was changed forever. A "next level up" communicating technique had been invented by Berners-Lee who, while working as a software consultant at CERN, developed the hypertext-based search approach that eventually led to the Hypertext Markup Language (HTML) that evolved into today's Internet browsers.

The dramatic revolution in mobile phones, begun with analog systems in the 1980s, sped ahead with digital modulation in the 1990s as operators became customer-driven by increased quality desires and ever greater capacity demands. Everyone appreciated that by using digital technology and techniques more, many more, facilities could be added. This fact became epitomized by the successful launch of products such as Nokia's 9000 "Communicator," that is, a PC-standard machine with 386-level power embodying a digital mobile phone with a text keyboard.

The user can send e-mail and faxes with this hand-held device because it links to the Internet via the local standard cellular network.

In the late 1990s, as the world rapidly approached the third millennium, digitization steadily expanded its horizons. The Internet was, of course, well in place with around forty million Web sites (and rising) being claimed in 1999, and Internet Protocol (IP) telephony was being seriously considered by increasing numbers of subscribers. Digital TV (DTV) was now very much on the scene (see Chapter 5) with interactive DTV (iDTV; also Web TV) gaining expanding acceptance. These types of developments drove up the long dormant interests in high-definition TV (HDTV).

Welcome to the gigahertz and terahertz high-speed digital millennium.

1.2 Increasingly interactive systems

Apart from the plain old telephone, most twentieth-century communications were one way at a time. Exceptions were audio and videoconferencing, but even here slow-motion video and other limitations (most notably cost) held growth back greatly.

Many CATV installations of the 1990s were largely fiber optic—at least as far as the street outside a group of subscribers. However, the fiber cables were what is known as multimode, which means that the bandwidth was limited to carrying only multichannel TV plus a number of telephone channels. Subscribers were generally content with the facility to watch, often on a pay-as-you-view near-video on demand (N-VOD) basis, whatever TV real-time programs or movies they wanted as well as in many cases to take advantage of the usually competitive telephone service. The service provider always built in a simple one-way keying option for the subscriber's use in program selection. As with the now all-digital systems, data could also be carried and hence computer (PC) modems could be connected.

By 1998 modem connection rates were as high as 51 Kbps in practice using such connections, and the existing fiber CATV networks could then also carry reasonably high-speed Internet connections. CATV modems, using special connections made available by some service providers, are already as high as 10 Mbps theoretically—2 Mbps in practice. These are downstream speeds. Upstream, 640 Kbps is achieved.

With ISDN, 64 Kbps is achieved as a local standard—but this is a dedicated network connection and it really predates the Internet.

Generally the "civils"—the trenching of the streets and laying cables—amount to around two-thirds of the total investment associated with any cabled project. This investment has already been committed with many urban fiber CATV networks, and at the appropriate time in the future, broadband single-mode fiber cables can be threaded through the existing plastic ducts relatively inexpensively. Such a move provides those subscribers who desire it with broadband connection facilities.

For example, HDTV with many more channels could be supplied, interactive TV could be made available, and high-speed Internet access could be facilitated. Instead of the 51 Kbps of 1998 the user could be connecting upstream at, say, 2 Mbps or more and this would mean the rapid transmission of graphics-rich files.

Almost all modern communications systems are digital and employ time-division multiplexing (TDM) or packet switching. Signals transported on broadband single-mode fiber cables are generally assembled on the "SONET" basis. SONET stands for synchronous optical network and the TDM sequences with this standard follow the synchronous digital hierarchy (SDH) scheme. SONET and SDH both allow bit rates up to at least 9.953 Gbps, which is the equivalent of having one single-mode fiber carrying 86,016 voice channels simultaneously! This is commercial; over one million simultaneous voice channels have been demonstrated in R&D labs.

The asynchronous transfer mode (ATM being the telecommunications engineers' abbreviation) represents a particularly important standard technique for assembling packets of information bits into a telecommunications bit stream. ATM is also sometimes known as "cell relay." Each ATM cell is a total of 53 octets long, where an octet contains 8 bits of information. A header section, just 5 octets long, precedes the 48-octet information field (or "payload") in every ATM cell. The overall structure is shown in Figure 1.5 and this technique is specified in the IEEE 802.6 as applying to metropolitan area networks (MANs), switched multimegabit digital services (SMDS), and the Broadband Integrated Services Digital Network (B-ISDN).

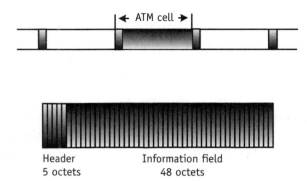

Figure 1.5   ATM cell. (*From:* Nellist, J.G., and E.M. Gilbert, *Understanding Modern Telecommunications and the Information Superhighway,* Norwood, MA: Artech House, Inc., 1999.)

The actual duration of an ATM cell varies according to the SDH level in which the cells are being transported. For example, in STM-1 the bit rate is 155.52 Mbps and therefore each bit has a duration of 6.43 ns. An octet therefore occupies eight times this, or 51.44 ns, and the complete 53-octet ATM cell must take up a 2.726 ms time slot. Obviously at higher SDH rates the duration is correspondingly reduced.

With these techniques available, digital compression approaches are feasible and have made great inroads into advanced digital telecoms. This is described in Chapter 5.

As we shall see later, new and ongoing developments are even more accommodating than the techniques described above.

Twentieth-century communications such as broadcast TV—terrestrial or satellite—remained essentially one-way and strictly receive-only. The basic concept of direct-to-home or direct broadcast by satellite (DTH or DBS) is illustrated in Figure 1.6. In this diagram a set-top box feeding into an HDTV set is shown, both of which are described in Chapter 5. However, in many cases traditionally a basic, even analog, TV is the receiving appliance for DTH or DBS satellite signals.

In North America satellite TV is provided by several companies including Primestar and USSB that are both now owned by DirecTV (itself owned by Hughes/RCA) and EchoStar. Dish reflector diameters vary considerably, being 90 cm for Primestar and just 45 cm in the cases

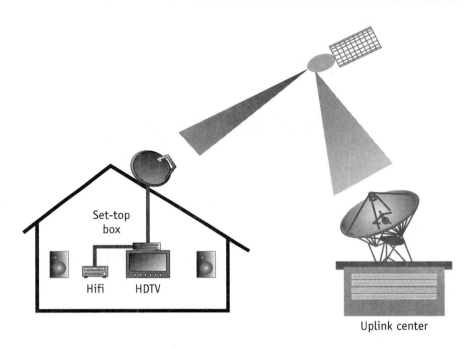

Figure 1.6 Direct-to-home (DTH) direct broadcast by satellite (DBS) system. (*From:* Nellist, J.G., and E.M. Gilbert, *Understanding Modern Telecommunications and the Information Superhighway,* Norwood, MA: Artech House, Inc., 1999.)

of DirecTV, EchoStar, and USSB. All programming is delivered from the uplink center and comprises a large selection of TV channels (e.g., live, local, news), movies, and music for all tastes at charges in the tens of dollars per month range.

Once thought of as a past opportunity, DBS enjoyed something of a revival in fortunes over the 1999–2000 time frame spurred on by digital TV, with its advantages of substantially improved signal-to-noise ratio and much greater programming flexibility including interactivity. The acquisition of e-mail and all other Internet usage was, however, also very much one way at a time with DTH and this is a significant disadvantage, although recently both DirecTV and EchoStar announced capabilities for Internet access.

One of the most exciting developments during the 1990s was the advent of mobile satellite systems (MSS). Iridium and Globalstar were the main first contenders in this "new space race," with Motorola and Loral as the lead contractors in the consortia. Although growth was slower than many hoped during the early period, notably in 1999, this approach led the way toward broadband MSS as an alternative to single-mode fiber—particularly in regions where cabling is completely uneconomic—and being mainly telephone connection systems two-way real-time conversations are made possible, including to and from remote regions of the planet. Only very low-speed connections are available, typically at just 4 Kbps.

Low earth orbit (LEO) satellites epitomized the early satellite TV of the 1960s and 1970s. However, the limitations of limited-period orbits and restricting technology soon meant that only geostationary orbiting spacecraft were used. Only advanced digital techniques and technologies have made the latest LEOs possible, but the economic viability of such operations, in particular narrowband LEOs like Iridium and Globalstar, has yet to be convincingly demonstrated.

Satellite trunks, mainly using the "Clarksian" (from the originator Arthur C. Clarke) constellations of three spacecraft in geostationary orbits (GEOs), also continue in importance. Their information-carrying capacities increase as on-board transponder capabilities improve and frequencies move ever higher, into Ka-band (26.5–36 GHz) and eventually V-band (46–56 GHz) also. Nellist and Gilbert describe many types of satellite systems, and broadband projects are discussed in Chapter 9 of this book.

Back on earth, broadband terrestrial wireless also continues to be of ever-growing importance. Bidirectional LMDS and MVDS systems are expanding in implementation, especially in regions where either satellite or fiber is uneconomic or impractical. Figure 1.7 shows a schematic of a point-to-multipoint configuration (more details are provided in Chapter 8). The millimeter-wave transceivers, complete with transmit-receive antennas (tightly focused MM-wave antennas A, B, and C), are mounted either on buildings or alternatively on dedicated units placed on hillsides or mountains. Line-of-sight trajectories are required between the main transceiver and A, B, and C, and the combination of "free-space," atmospheric, and precipitation attenuation must be accounted for in the link design. Operating bands are usually 25–31 GHz and all forms of attenuation increase as frequencies shift into these millimeter-wave bands.

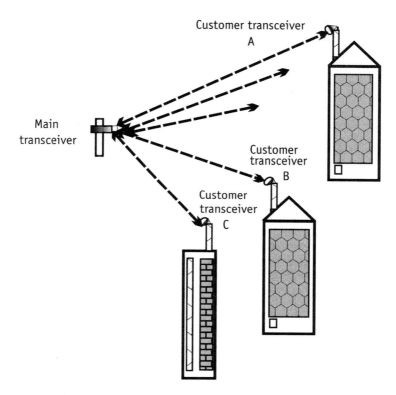

Figure 1.7   Point-to-multipoint communications.

While important technologies such as PCS and especially GSM remain in vogue for millions of mobile phone users and service providers globally, new "third-generation" (3G) specifications and associated products are also entering the markets. Two considerations that are of significance here are the Universal Mobile Telecommunications Systems (UMTS) specifications and wideband code-division multiple access (W-CDMA). This is also sometimes known as B-CDMA, broadband code-division multiple access, which is well covered by Brodsky in *Wireless: The Revolution in Personal Telecommunications*. Standards associated with these new developments are coming from the IEEE, ETSI (European), IMT, and other bodies. IMT 2000 represents just one group of such standards. New spectrum allocations are being made available by the World Administrative Radio Conference (WARC) in the year 2000.

An example of a multimedia "3G concept phone" being developed by Nokia is indicated in Figure 1.8. Important characteristics include the ability to operate using almost any of the standards available globally, that is, true global roaming capabilities. Such products also enable Internet access from the mobile. All this is made possible by implementing advanced single-chip technologies and highly adaptable antennas. Products of this general type are being dubbed "millennium phones" in North America.

These broadband satellite systems and terrestrial wireless services represent additional on-ramp options for the global information superhighway, complementing optical systems.

1.3 Why gigahertz wireless bearers?

Most conventional broadcast radios operate at carrier (bearer) frequencies anywhere in the approximate range from several hundred kilohertz (kHz) up to several thousand MHz (i.e., GHz). However, the familiar bands include long-wave and medium-wave (both kHz) and also VHF that extends from several tens of MHz to 300 MHz.

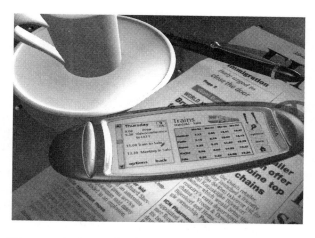

Figure 1.8   3G concept phone. (Artist's impression courtesy of Nokia Mobile Phones, 1999.)

Mobile (cellular) communications use bands extending from around 400 MHz to at least 2,000 MHz (2 GHz) for personal communications service (PCS) in North America. This also applies to DECT cordless in Europe or the personal handyphone system (PHS) in Japan. The latter carrier frequency, 2,000 MHz or 2 GHz, is within the lower microwave ranges and the antennas associated with most systems of this type are quite directional. This means that the transmitted signal energy is mainly concentrated into a spatially well-defined beam.

In contrast, an antenna transmitting on a carrier frequency of only about 500 kHz or even 200 MHz is almost omnidirectional, that is, the signals tend to propagate over wide surrounding areas. GSM mobile bands, in the 900–1,000-MHz range, naturally become associated with quite directional antennas. So your mobile phone antenna "looks" strongly in one principal direction wherever you are—at any particular location.

The types of possible antennas that can be implemented are very important determinants toward the directionality of transmitted and received signals. Paraboloidal dish reflectors, or similar antennas used in microwave, millimeter-wave, and satellite systems, lead to highly directional beams, and the general situation is illustrated in Figure 1.9.

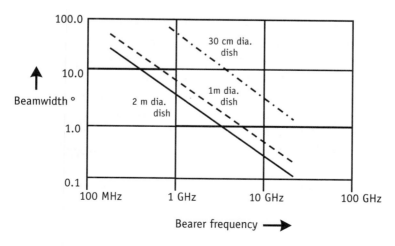

Figure 1.9   Antenna beamwidth versus frequency.

With antennas the so-called beamwidth is the main measure of directionality. The term *beamwidth* is really a misnomer because we actually mean the enclosed angle within the beam, measured at the half-power points. This beamwidth is a function of both frequency and dish diameter—decreasing if either or both of these parameters is increased. In Figure 1.9 it is assumed for simplicity that the same diameter (always 2m) antenna dish reflector applies in each case, for three different frequencies increasing by factors of 10: 200 MHz, 2 GHz, and 20 GHz. The beamwidth then progressively decreases by the same factors of 10: from a wide 38° at 200 MHz, down to 3.8° at 2 GHz, and finally right down to a very narrow 0.38° at 20 GHz. These trends are summarized in Figure 1.10, in which three dish diameters are the parameter, that is, 30 cm, 1m, and 2m.

This clearly shows that the beamwidth decreases linearly with frequency under conditions that are otherwise fixed. When the antenna is not of the dish reflector type, an equivalence applies and beamwidths still decrease with frequency.

It should be clear now that the general idea of this beamwidth is similar to those of light torches or car headlamps with their concentrated beams. The more focused the lens—the more concentrated the beam of light becomes—and the narrower the beamwidth is.

If we keep all other parameters constant and just continually raise the frequency then the beam becomes ever more concentrated—the beamwidth continually decreases. At lightwave frequencies, well into the terahertz ranges (1 THz = 1,000 GHz), beamwidths drop to small fractions of a degree and the beams are extremely tightly focused. This is why laser beams transmitted from earth and directed at the moon's surface are still only spread out to the extent of about 1 ft (around 0.3m). Yet such a beam will have traveled 239,000 miles or well over one third of a million kilometers. A microwave beam sent the same distance would cover all or most of the moon. Tight pencil beams provide for frequency reuse and improved information security.

The reception coverage area in the vicinity of the target, the region for which the signal is destined, is usually termed the *footprint* and this term is widely used in satellite communications. Similar considerations also apply in cellular (mobile) communications where the footprint is a single cell.

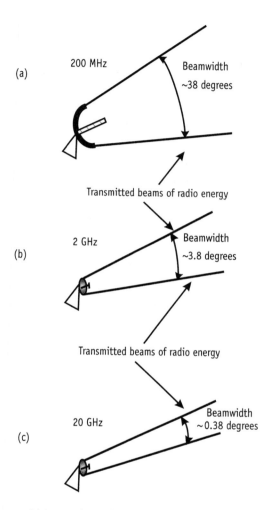

Figure 1.10 Beamwidths at three frequencies.

It is significant that the beams of radiation become particularly narrow as we shift toward millimeter-waves, which means that the pointing control for the antennas becomes increasingly critical due to the finely focused beams.

Directionality or narrow beams are, however, decidedly not the only reason for using gigahertz bearers. The other reason concerns available *bandwidth*. Bandwidth here means frequency bandwidth, that is, the range of frequencies needed to transmit at a given bit rate. Figure 1.11(a–c) illustrates typical bandwidths, which are often quoted in percentage terms. These are generally termed either *frequency-domain charts* or more simply *spectra*.

Bit rates in the high hundreds of megabits per second and into gigabits per second are needed to cope with the ever increasing telecommunications traffic globally, interactive multimedia transmission, and the global information superhighway.

The transmission of digital signals with bit rates in the region of 622 Mbps or above (STM-4 level in the SDH hierarchy) requires at least comparable frequency bandwidths. Since it is impossible to have such

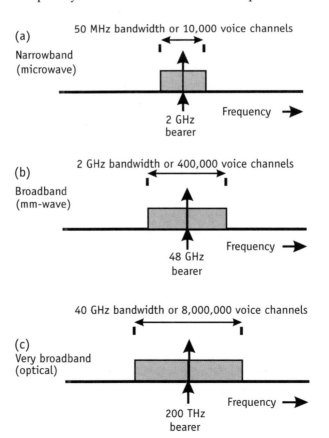

Figure 1.11   Three examples of relative bandwidths.

bandwidths with megahertz bearers, there is an absolute necessity to shift at least to gigahertz.

In Figure 1.11(a), the bearer frequency is 2 GHz and carries an information bandwidth of 50 MHz. This translates to a percentage bandwidth of 2.5%, which is definitely considered narrowband. The next spectrum shown in Figure 1.11(b) represents a 2-GHz bandwidth symmetrically distributed around a 48-GHz bearer and, using the same type of calculation, the percentage bandwidth in this case is just over 4%, which would be considered relatively broadband. Finally, Figure 1.11(c) shows that a very broadband transmission spectrum—with a 40-GHz bandwidth on a 200-THz optical bearer—is interpreted as 0.02 %. Although this is a very narrow bandwidth in percentage terms, the available bandwidth as far as information transmission is concerned is actually large. A total of 8 million simultaneous phone conversations could be transported within this band.

The 622-Mbps signal could be carried by the 48-GHz bearer within the available 2-GHz band. However, higher bit rates require still more bandwidth and the pressure is on to shift to higher bearer frequencies. This shift comes at a price, however, because of millimeter-wave component scarcity and immaturity which leads to relatively high costs, and also limitations due to signal attenuation as it passes through the earth's atmosphere. This signal attenuation is a strong function of frequency and dictates, together with World Radio Conference (WRC) and other local regulations, the choice of frequencies for any radio-based system.

Microwave and millimeter-wave signal attenuation is presented and its implications are discussed in Chapter 2. Many new and proposed systems will operate at millimeter-wave frequencies where attenuation can become high (e.g., V-band which is 46–56 GHz), and therefore this signal attenuation feature is particularly significant. Apart from free-space loss and atmospheric attenuation, rainfall is also a serious issue at these high frequencies and 8 dB of additional loss through heavy rain is not uncommon.

1.4 Cabled systems and terahertz transmission

Until the advent of fiber optics, all cabled systems were of copper, either twisted wire pairs or coaxial for higher frequencies. Locally, for the final drop to the subscriber's premises, whether residential or business, copper has been and often still is the name of the game.

By the early years of the third millennium a vast heritage of copper final drop connections were in place—notably in the G8 advanced economies. Again, digital techniques and clever technologies such as ADSL have enabled conventional copper cables to cope with higher bit rates than hitherto believed possible. For example, ADSL enables up to 6 Mbps signal transmission rates on copper twisted pair lines to a distance limit of approximately 2 km, which is suitable for the local loop.

Technologies such as ADSL work well enough until:

- Higher bit rates or bandwidths are required, reaching over similar distances.

- More modest bit rates are involved (e.g., 6 Mbps) but reach over longer distances.

- Both of the above.

For interactive multimedia transmission on the global information superhighway much wider bandwidths are demanded at the point of use, that is, at the subscriber. ADSL-like technologies no longer apply and single-mode fiber optics are essential.

The late twentieth century saw CATV networks increasingly implementing fiber optics as described above. But these were narrow-band systems compared with the capabilities of single mode fiber, and network providers have to reinstall this future-proofing type of cable into the existing ducting.

As remarked earlier, intelligent lightwave networks using DWDM are now state-of-the-art for trunk cabled telecommunications. Carrier frequencies, the "bearers" with this technology, are in the 200-THz class and available bandwidths are enormous even when compared with millimeter-wave radio [see Figure 1.11(c)].

Some of the types of services that require gigabit per second-to-terabit per second transmission and involve bandwidths in the gigahertz ranges are summarized in Figure 1.12.

Total aggregate bit rates on optical systems have increased from over one terabit per second in the late 1990s to many terabits per second now.

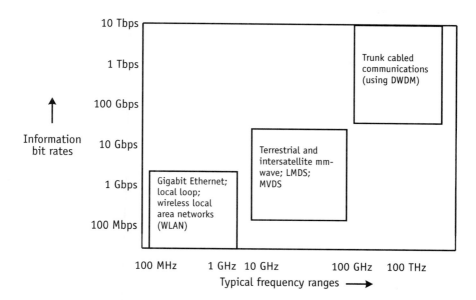

Figure 1.12 Systems applications: bit rates and frequency ranges.

It is a sobering thought that this vast information transfer rate all occurs in a glass fiber core only a few microns in diameter. Signal processing technologies have shifted inevitably from monolithic microwave integrated circuits (MMICs) to monolithic photonic integrated modules (MPIMs).

Major players in this sector include Fujitsu, Lucent Technologies, NEC, Nortel, and Pirelli, who supply products such as DWDM modules, fiber-optic amplifiers, lasers, and various time-division multiplexing (MUX) subsystems.

Another advantage of fiber-optic transmission is that microwave and millimeter-wave signals can be used to modulate the laser transmitters and hence ultimately transport the signal optically. A schematic view of this type of subsystem is indicated in Figure 1.13. The (usually) digitally modulated electrical signal, using a microwave or millimeter-wave bearer, is inputted to the semiconductor laser module. This module often comprises a distributed feedback (DFB) laser complete with DC power supply, electronic driver, and interconnections including provision for the optical fiber output.

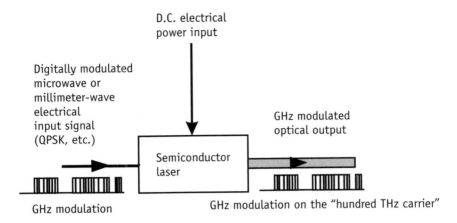

Figure 1.13   Microwave or millimeter-wave modulation of a laser (for very high-speed optical transmission).

The output signal format, frequently on a "bearer" at around 200 THz, is then either amplitude modulated or coherently modulated with the original microwave/millimeter-wave electrical input signal. In this case, *amplitude modulation* simply means that the power level of the photon stream is varied in accordance with the original electrical power level. Coherent modulation is considerably more sophisticated because here the relative phase or time-dependence of the optical signal is varied—and this is demanding at 200 THz.

In this way signals originating from wireless sources, including mobiles, can be rapidly and conveniently transferred onto the optical backbone.

1.5 The available component technologies

Fully solid state component implementation, mostly using integrated circuits, fits the bill for most systems functions. At microwave frequencies MMICs are extensively deployed, for the low-power sections of millimeter-wave systems MWMICs (millimeter-wave microwave integrated circuits) are available and, as mentioned above, MPIMs are being incorporated into photonic systems. An example of an MMIC, fabricated in gallium arsenide (GaAs), is shown in Figure 1.14. In common with

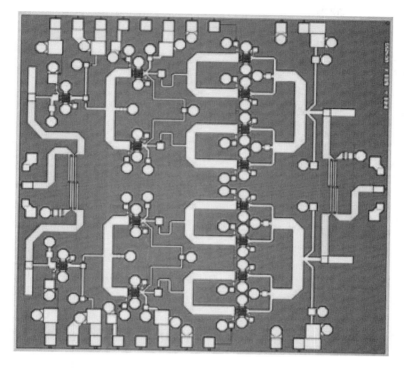

Figure 1.14 A GaAs MMIC chip. (Courtesy of Marconi Materials Technology.)

all ICs, these types of chips start their lives as rectangular pieces of semiconductor a few millimeters square and around one-fifth of 1 mm in thickness. Using micromanipulation techniques, often at least semi-automated, the chips are mounted into carriers and encapsulated for circuit handling and environmental protection.

MPIMs are described in Chapter 3.

In an overall sense we are already well into what could be described as the "nano age." The trend from "micro" to "nano" ages is most clearly seen by examining how computer processor chips have progressed since the 286 and this is illustrated in Figure 1.15.

The march of progress has been dramatic and inexorable. When the personal computer was invented at Xerox in the late 1970s, 100,000-transistor chips with submegahertz clock rates were state-of-the-art and memory remained at a distinct premium. If you found a RAM

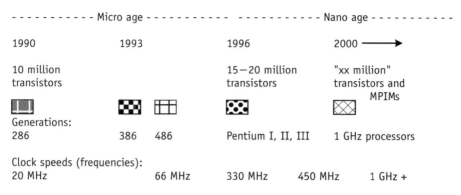

Figure 1.15   From micro age to nano age—from megahertz to gigahertz processors.

chip with 15K (yes, kilobytes) of memory you had done very well and disks were things you inserted into relatively large and expensive minicomputers.

Now, into the twenty-first century, processor chips clocked at above 1 GHz and with over 30 million transistors have become the state-of-the-art; 64- and 256-MB RAM chips are available and hard disk drives regularly enable up to 10 GB of memory capacity. In the nano age clock rates are very fast, processors are powerful, and memory capacity is almost unlimited.

With the now intense convergence between all things "computer" and everything associated with telecommunications, all these trends are vital considerations in the context of developing technologies.

Sometimes future system concepts are put forward to test the market and occasionally to book early for regulatory permission. Frequently, future system plans are launched to an unsuspecting public, many of whom marvel at the tremendous capabilities specified for the systems. Some of these same people wonder just how the technology is going to be developed to support the systems concepts and how the economics may work out.

This scenario is not uncommon with some millimeter-wave and satellite systems at their earliest stages. While suitable component technologies may be in advanced development, it is often many years before such components can be remotely considered mature—and then there is the question of cost.

Although the electronics world understandably expects solid state technology to provide all the answers for every aspect of any system, this is far from the case in most millimeter-wave and satellite systems. In one word, the problem leading to a continued need for electron tubes is *power*. Continuous power levels in the tens-to-hundreds of watts can only be made available using semiconductors when the frequencies involved are in the lower microwave ranges. At higher frequencies electron tubes must still be used and the traveling wave tube (TWT) is the most important example.

Transmitters for satellite earth stations and for some MVDS systems frequently use TWTs and this scenario is likely to persist for another decade at least.

Other significant technological advances include aspects of filter structures, antennas, and the burgeoning area of software.

Frequency band filters are increasingly densely integrated, and hybrid microwave (and millimeter-wave) circuit modules provide the desired functions in many instances. However, for demanding specifications or high power applications, waveguide or coaxial filters still need to be implemented. One advantage, however, is that as frequencies increase into the millimeter-wave bands the physical dimensions of such filters and other "passive" modules decrease.

Up to frequencies of the order of 10 GHz, a novel and remarkable technology known as surface acoustic wave (SAW) continues to make great inroads. With this technology the electrical signal is converted into an acoustic vibratory replica, that is, corresponding mechanical (acoustic) vibrations traveling just beneath the surface of a quartz or lithium niobate chip. These chips have dimensions that are just a little larger than those of the semiconductor ICs, and tiny etched metallic patterns on their surfaces provide the filtering function. SAW filters and other related devices provide exceptional selectivity and stability.

SAW devices are used as highly selective frequency filters in many systems, including those using frequency-division multiplexing. Very low in-band loss and sharp band edges with strong out-of-band signal rejection are important characteristics of SAW filters used in gigahertz systems, but this technology may never be able to reach the important Ku- and Ka-band frequencies.

Antennas represent another technology area where substantial advances continue to be made. Although the familiar parabolic dish reflectors will still be seen on many installations, new configurations are also making considerable inroads for many systems. These include flat-plate designs where a metallic array is etched onto a plastic sheet, modular lensed antennas for millimeter-wave transceivers, active arrays, and phased arrays.

Planar metallic arrays are an economical method of forming antennas, provided the application is not too demanding in terms of beamwidth and sidelobes. This applies to many instances, including several of the systems described in later chapters here. Modular lensed antennas, notable for their important applications in millimeter-wave transceivers, are described in Chapter 8 where stratospheric and terrestrial systems are considered.

The technological advances combined with unit price decline have conspired to make active microwave and millimeter-wave chips (MMICs) serious contenders for implementing directly into antenna arrays. This means that antenna arrays may have transceivers embedded within their structures, enabling independent control over each element's characteristics. While certain elements remain active transmitting, other selected elements may be switched "off," that is, not transmitting. This provides for a high degree of flexibility in both transmission and reception. Different elements may also be fed with the main signal in differing phases to obtain a similar overall resultant pattern—what is termed a p*hased array*.

In the schematic shown in Figure 1.16, the filtered and multiplexed signal is fed into the structure from the left hand side. DC power also has to be supplied to each of the MMICs which are selectively driven with predetermined portions of the signal.

Examples of phased arrays include some of the antennas used for the Iridium mobile satellite system and many defense systems. Alcatel Espace is developing its "SkyBridge" broadband satellite system which implements an active array with a notably large number of MMICs and 10,000 solid state power amplifiers (SSPAs). More details on phased arrays and the SkyBridge active array are provided in Chapters 4 and 9.

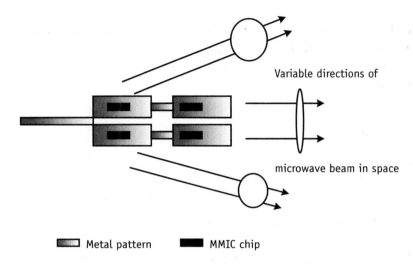

Metal pattern   MMIC chip

Figure 1.16   Conceptual schematic of an active or phased array planar antenna.

1.6 The ongoing communications revolution

Without any shadow of a doubt the world is continuing to enjoy the fruits of a radical revolution in communications. During the twentieth century major advances were seen in the digitization of most forms of communications including both fixed and, later, mobile networks. Increasingly powerful enabling technologies such as fiber optics and satellite transmission, supported synergistically by microminiaturized electronics (integrated circuits), transformed communications.

System link bandwidths were capable of handling hundreds of thousands of simultaneous telephone calls or data at bit rates up to several gigabits per second. Many people could be forgiven for imagining that this was practically the ultimate or at least that channel capacities would simply go on increasing—but not much else. Then, originally at CERN in the late 1980s, a guy named Tim Berners-Lee invented HTML—and the world of telecommunications was changed forever with what inevitably followed, namely the Internet.

Meanwhile very high-speed broadband systems using fiber optics, satellites, and terrestrial configurations were planned and these would definitely be making a dramatic improvement in terms of getting on to the Internet. Early in the twenty-first century the implementation of broadband systems of several types will mean that the "world wide wait" will at last be transformed into a true "world wide wizard."

Read on to find out much more about these key enabling technologies.

Select bibliography

Brodsky, I., *Wireless: The Revolution in Personal Telecommunications*, Norwood, MA: Artech House, Inc., 1995.

"FT Telecoms," *Financial Times* Survey on Telecommunications, March 18, 1999.

Nellist, J.G., and E.M. Gilbert, *Understanding Modern Telecommunications and the Information Superhighway*, Norwood, MA: Artech House, Inc., 1999.

Willner, A.E., "Mining the Optical Bandwidth for a Terabit per Second," *IEEE Spectrum*, April 1997, pp. 32–41.

2

The Hunger for Bandwidth

2.1 Why ever-more bandwidth?

In some instances the need for relatively large bandwidths is almost self-evident. While speech occupies only a few kilohertz of bandwidth, transmitting just one broadcast TV channel demands about 6 MHz. This is three orders of magnitude more than a single basic (analog) telephone channel. Similarly the transmission of data begins with single sources operating at perhaps 10–100 Kbps and data compression techniques are regularly used to reduce bandwidth demands.

The question is therefore acutely valid: Why is bandwidth an ever-more scarce "commodity" and why is there the constant pressure for more bandwidth and indeed broadband systems?

The perfectly valid question must be asked: since local bit rates are in the kilobits-per-second class—such as the 51 Kbps used in practice for many 1999–2000 Internet connections—why on earth are megabits per second, let alone gigabits per second, needed at all?

Part of the answer lies in the need to efficiently cram as many channels as possible into single bearers, that is, on single higher frequency carriers. Other factors include:

- Bringing the Internet out of the "world wide wait" and into a more effective high-speed World Wide Web;

- Increasing demand for broadband interactive communications.

The cramming of multitudes of individual channels together into a form of superchannel is termed *multiplexing*. So there is almost always the fundamental need to multiplex in order to efficiently use the common telecommunications fabric.

If everything remained at or around 51 Kbps or 64 Kbps, then all we could possibly have would be literally billions or even trillions of dedicated lines (all copper) interconnecting each and every possible subscriber on the planet! Every new subscriber would have to be supplied with a new set of copper cables—some extremely lengthy to the extent of thousands of miles or kilometers! This is obviously ridiculously clumsy, expensive, and impractical—as well as the environmental concern stemming from planet Earth being covered with a copper cable network so vast that there would be hardly enough room left to live in.

A much more efficient approach, in fact the only practical and economic approach, is to provide a common (if complex) network infrastructure that all subscribers can share. This is the telecommunications infrastructure—vast and always growing more and more complex and intertwined.

There are basically two approaches to multiplexing: frequency-division multiplexing (FDM) and time-division multiplexing (TDM). The basic principles of each approach are indicated in Figure 2.1. FDM has the longest history because telecommunications and radio people have traditionally thought in terms of frequencies more readily than in terms of time intervals. With this approach a relatively wide frequency spectrum is carved up into "chunks" of much smaller bands of frequencies—shown as channels C1, C2 ... Cn in Figure 2.1. At the most basic telephone level, the frequencies could be in the kilohertz through hundreds of kilohertz ranges. As the multiplexing proceeds to higher levels so the channel frequencies rise into the megahertz (radio), then gigahertz

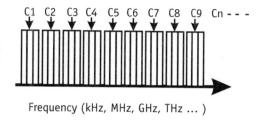

Frequency (kHz, MHz, GHz, THz ...)

Frequency-division multiplexed (FDM) channels

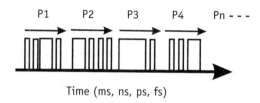

Time (ms, ns, ps, fs)

Time-division multiplexed (TDM) frames or packets

Figure 2.1 FDM and TDM compared.

(microwave and millimeter-wave), and finally the terahertz levels. The latter are associated with optical systems.

Consider, for example, a simple situation where the channel bandwidth for the data being transmitted from your computer modem is, say, 80 kHz. At some later point your message could well be going over a satellite link where the lowest frequency in your channel is 22 GHz and then the highest frequency is 22.000080 GHz. Your channel reservation could be anywhere in a very short or a very long communications sequence. At all times and places your message is fully identified in terms of sender and recipient(s). Obviously there exist a vast number of additional channels for other users and "guard bands" must be inserted between adjacent information-bearing bands to prevent overlap.

Although FDM is still used in many communications systems, the digitization of messages and their processing means that time-oriented rather than frequency-oriented techniques are more appropriate and so TDM is extensively employed. With the TDM approach the elements of the messages are allocated time intervals termed *frames, packets,* or *cells* and the basic concept is shown in Figure 2.1 (lower diagram). Here P1, P2, P3 ... Pn are the frames, packets, or cells and there is some analogy with FDM in that the actual time intervals extend downwards as the TDM level increases.

At the most basic telephone or ex-modem data transmission level, the time intervals could be in the milliseconds (ms) to microseconds ranges. As the multiplexing proceeds to higher levels so the channel time intervals decrease down to the nanosecond (ns), then picosecond (ps), and finally the extremely small femtosecond (fs) intervals. Picoseconds and femtoseconds are associated mainly with coherent optical systems and this is touched upon in Chapter 3.

Assuming a 64-Kbps encoding rate and the use of (as standard) pulse code modulation (PCM) for speech, the basic synchronous digital hierarchy (SDH) multiplexing process is shown in Figure 2.2. The internal complexities of the individual elements of each processor are omitted from Figure 2.2 for clarity. At the top left a set of 24 telephone input channels is indicated and each enters the PCM channel bank. The output bit stream from this channel bank is then slightly higher than 1,536 Kbps directly from the inputs and is in fact 1.544 Mbps. Digital data streams for computer communications also enter the multiplexer.

The combined PCM and digital data streams are fed into an SDH level-1 multiplexer (MUX) and the output is then at the STM-1 level, namely 155.52 Mbps. There can be several digital data streams and the entire process at this level may be duplicated many times—a second level-1 arrangement is shown in Figure 2.2. All level-1 processors generally feed into a level-4 SDH MUX to produce the STM-4 output at 622.08 Mbps and the entire process may itself be replicated up to at least STM-64 (9.953 Gbps).

With these types of arrangements a fundamental consequence is the need for multiplexing at the transmitters and demultiplexing at the receivers. The 64-Kbps digital signal travels along the subscriber's local cable until it is multiplexed either at the exchange or at the local CATV operator's curbside box. Everyone else's 64-Kbps digital signal is treated in precisely the same way and an assembly of typically hundreds of "time packet" multiplexed signals is built up in this manner.

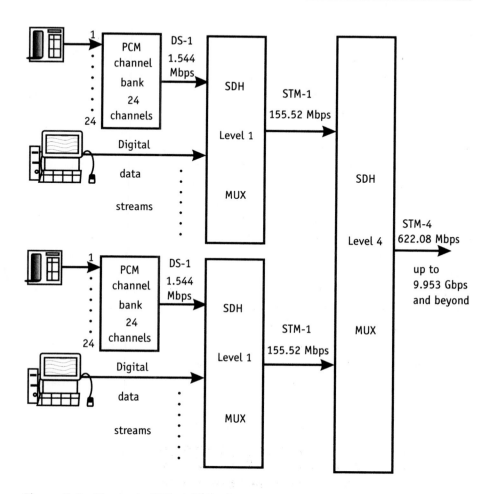

Figure 2.2 The basic SDH multiplexing process.

Alternatively, in a large organization such as corporate offices or a university, all the locally originated digital signals are sent to a central unit where all the organization's multiplexing and demultiplexing takes place. In such circumstances some individuals may be privileged in that they can sometimes have high-speed dedicated lines that may enable them to access the Internet, or other facilities, at speeds like 2 Mbps.

The idea is quite simple: "packets" of data, literally data as such or bit sequences derived from phone conversation signals, are assembled into longer and longer sequences over time. Specific packets belong to specific subscribers and are appropriately addressed (tagged) as such. With sophisticated addressing it is not necessary for adjacent packets to belong to the same subscriber because all the packets are sorted and directed accurately by the system.

The concept is illustrated in Figure 2.3 where a sequence of three ATM cells is shown. As described in Chapter 1, each ATM cell comprises the initial 5-octet header followed by the 48-octet information-bearing portion. The 5-octet header includes routing (tagging) information. Parts of your signal may, for example, be contained in cells 1 and 3, while cell 2 contains information originated by someone many hours distant from your location. Appropriate addresses of senders and destinations are contained in the headers that route the messages that are finally all reassembled at the ends of the route.

However, time itself is the critical parameter and if we retained the (slow) subscriber bit rates for the multiplexed streams, the overall time intervals would become intolerably prolonged. To grasp this fully consider just one hundred subscribers each having packet lengths (i.e., time slots) containing 424 bits, which is fairly standard. Transferred at the typical modem rate of 51 Kbps the total duration for a packet is 8.3 ms.

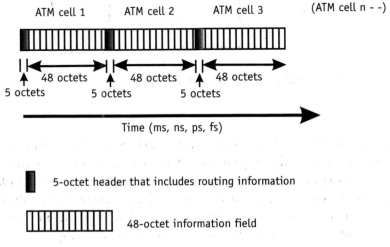

Figure 2.3   Information stream as sequence of ATM cells.

Next assume for simplicity that each subscriber needs to occupy 20 information packets or cells—which amounts to 166 ms. Across all one hundred subscribers the time required is clearly 16.6 seconds which represents an intolerable delay such that phone conversations, for example, would be impossible. (Although the preceding discussion relates to data transmission the same basic principle applies to digitized speech.)

The problem is even more critical with bandwidth-hungry applications such as the Internet, videoconferencing, multimedia, and entertainment TV.

To overcome this delay problem the bit rate must be accelerated so that, regardless of the distances involved, communications appear immediate to the subscribers. This is one reason why bit rates have to rapidly climb into the megabits per second, hundreds of megabits per second, and, in many instances, on trunk routes or for other purposes, into the gigabits per second ranges.

Individual subscribers operate with relatively slow signals, some kilobits per second, or perhaps 1 Mbps or somewhat more, but on the trunk routes all their information is literally racing along at very much faster rates.

Channel capacities may be considered in frequency (and bandwidth) terms or alternatively in the time domain, and Figure 2.4 shows the comparisons. The top portion of this chart provides a general view of the electromagnetic spectrum ranging from DC (i.e., 0 Hz) through radio, microwave, millimeter-wave, and on to optical or photonic frequencies in the terahertz bands. Engineers tend to term this the *frequency domain*. Sections of the bands shown separated by increasingly close verticals indicate approximately "scaled" bandwidths, with more and more signal channels becoming available as the frequency rises.

The lower portion of the chart is intended to approximately match the upper spectrum area and refers to bit rates or "information transfer rates." Instead of frequencies, we now operate in the time domain and the units are bits per second (bps). Waveforms of the digital information streams are now indicated. Bit rates range upwards from kilobits per second, through megabits per second and gigabits per second, to the terabits per second of which optical systems are capable of carrying. Kilobits per second tend to go together with kilohertz and megahertz, whereas

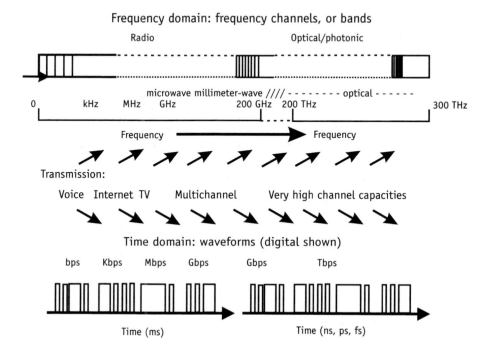

Figure 2.4 Channel transmission viewed in the frequency and time domains.

megabits per second through gigabits per second are associated with gigahertz, and so forth.

Increasingly bandwidth-hungry applications go from left to right across this entire chart, starting with voice and low-bandwidth Internet connections and proceeding through to high-level multiplexed applications. This is summarized in Figure 2.5 where the bandwidth demand is plotted to a base of the increasingly demanding applications. Voice (POTS) is the least demanding, and in contrast entertainment TV (particularly iDTV and HDTV) is the most demanding of bandwidth. It should be noted that this chart is generated on a per channel basis—that is, individual single channels are considered here.

Signals associated with all these applications are frequently transmitted by wireless means, and prior to this, high-level multiplexing is applied in a manner similar to that shown in Figure 2.2. The final bit rates,

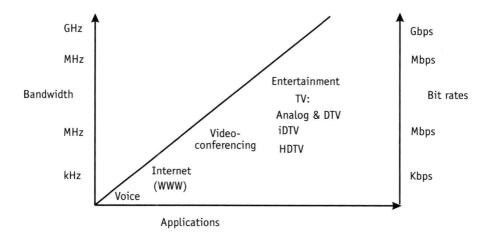

Figure 2.5   Bandwidths and bit rates required for various communications applications.

often hundreds of megabits per second and higher, require substantial bandwidths and the total bandwidth available for such signals is limited in all cases.

What is often disregarded, at least until an excessively late point in considerations and plans for new systems, is that:

> The available bandwidth in radio (microwave/millimeter-wave) and optical systems depends markedly upon the region of the electromagnetic spectrum in which the system is desired to operate.

Further details regarding this "wireless canon" are provided later in this chapter.

2.2 Future bandwidth resources

There is often confusion between *bandwidth* and *bit rate* and the true meaning of these terms. The foregoing discussions should clarify just what bandwidth and what bit rate amount to as well as the fact that as bit rate demands increase so indeed the required bandwidth for transmission also increases.

So here is an absolutely key question: If you double the bit rate demanded, do you also simply double the bandwidth required for transmission? Unfortunately, the answer is negative. Things are just not that simple—they hardly ever are! This is because, aside from bandwidth and bit rate, there are two further important considerations:

1. The modulation scheme chosen;

2. The relative amount of noise power in the channel.

With digital communications systems several modulation schemes are available: phase shift keying (PSK), quadrature phase shift keying (QPSK), 128-TCM (trellis-coded; described later), and so forth. Even with constant noise in all instances the bandwidth required will always differ according to the choice of modulation scheme. In many instances the selection of modulation scheme is at least partly contingent upon its bandwidth efficiency, that is, the maximum bit rate practically achievable using a given bandwidth.

Unfortunately it is not only our wanted signals that are present throughout transmission. Instead we have to contend with pervasive unwanted disturbances, generically termed *noise*, that adversely influence all electrical and optical systems. The average power arising from this noise can often itself be broadband, therefore affecting all signals across the spectrum. A major objective in communications systems design and implementation is to be able to distinguish the desired signals from the background noise, and then receive the messages accurately.

Many decades ago Claude Shannon discovered that, even for basic digital (binary) modulation, bit rates could never simply increase proportionately with bandwidth. The effects of this background noise in the communication channels also clearly impact the bandwidth required to transport a given signal at a known bit rate.

The upper diagram of Figure 2.6 provides a concept of signal and noise. It is very important to appreciate that the total indicated radio bandwidth actually contains typically several thousand speech channels each only occupying a few kilohertz of bandwidth.

In the PCS example shown in Figure 2.6 the signal is distributed across the 30-MHz bandwidth of the channel—but so also is the noise. However, because of the random nature of the noise, increasing the

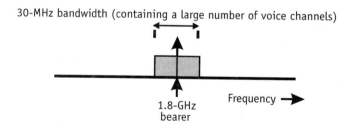

30-MHz bandwidth (containing a large number of voice channels)

1.8-GHz bearer

Frequency ➤

Spectrum of the 30-MHz total bandwidth on the 1.8-GHz (PCS) bearer frequency

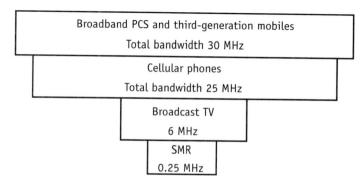

Broadband PCS and third-generation mobiles

Total bandwidth 30 MHz

Cellular phones

Total bandwidth 25 MHz

Broadcast TV

6 MHz

SMR

0.25 MHz

Figure 2.6 Bandwidths for a selection of well-known services.

bandwidth may do little for the signal and yet increase the noise considerably, making communications more difficult with most systems. Engineers talk in terms of the ratio between the signal power and the average noise power in the channels.

In order to successfully transport signals, in many instances the signal power has at least (on average) to exceed the average noise power. However, with some innovative systems the signal can be literally buried within the noise and yet accurate reception remains possible. This approach is referred to as *spread-spectrum* and very wide bandwidths are used. Brodsky has an excellent nontechnical treatment of spread-spectrum techniques.

The concept of bandwidth efficiency, referred to above, is extremely important. Additionally, the choice of carrier frequency is also highly significant.

Most wireless systems, particularly cellular mobiles and cordless phones, utilize carrier frequencies up to a maximum of around 2 GHz (2,000 MHz). This maximum is dictated by a combination of technical and economic parameters. In Chapter 1 we noted that as the carrier (or bearer) frequency rises so also the beamwidth decreases, that is, antennas become increasingly directional. But in the absence of sophisticated technology many services demand precisely the opposite—users will be anywhere but exactly in this or that direction with respect to both the antennas and base stations, and all-around or "omnidirectionality" is really needed. In summary, keeping frequencies relatively low (less than about 2 GHz):

- Provides reasonable "all round" access (i.e., omnidirectional);
- Keeps costs and ultimately unit prices down for the end-users.

However, bandwidth is very restricted and truly broadband operation is not possible. Figure 2.6 shows the 30-MHz bandwidth (upper diagram) and a comparison between four practical instances in the lower diagram. The 30-MHz bandwidth available with PCS supports the major aim of the service, which is to provide for a substantial number of voice channels, but not much else and certainly not full Internet connectivity (not more than e-mailing anyway).

However, when compared with several other common services, PCS does have a relatively wide total available radio bandwidth. This is clear from the lower diagram in Figure 2.6 where specialized mobile radio (SMR), broadcast TV, earlier cellular phones, and PCS are all compared. Traditionally broadcast TV has been assumed to have a wide required bandwidth and yet this is only one-fifth of the PCS availability. We shall see that new services generally have even greater bandwidths than any shown in Figure 2.6.

The main frequency standard for radio-based wireless LANs is the ISM (industrial, scientific, and medical) frequency of 2.45 GHz, which is insufficient for supporting anything like truly broadband signals. In this instance the now considerable directionality is much less important than with mobiles.

By the end of the twentieth century a radio bandwidth crisis had developed. Right across the radio spectrum, certainly from kilohertz frequencies up through microwave, available carrier frequencies and available bandwidths around them became so congested that many techniques had to be invented to circumvent the problem, as far as possible. Major examples of such innovations include adopting bandwidth-efficient modulation techniques and implementing "xDSL" approaches—particularly asymmetric digital subscriber line (ADSL).

None of these techniques was without its disadvantages, and compromises were necessary. This is especially noticeable in the case of modulation techniques where special coding arrangements providing extremely low bit error rates in reception have led to methods including trellis-coding. There is, for example, trellis-coded quarternary phase-shift-keying (TC8PSK).

The scarcity of radio bandwidth has already led to the concept of "bandwidth trading" and this has gained considerable momentum. With this concept, when a new user requires $x\%$ of bandwidth then, instead of applying to the relevant regulatory organization, the new user can purchase this bandwidth from another previous user who no longer requires the particular band. Of course, regulations demand that the new user check with the appropriate regulatory authority first. This process enables the marketplace to naturally tend to fill all or most gaps in the entire commercially available radio spectrum in a dynamic manner.

In several applications the basic solution was to shift upward into millimeter-wave, and the first "green shoots" of this appeared during the final two years of the twentieth century. For mobile communications the use of higher frequencies necessitates automatic tracking, fast acting, and highly sophisticated technology so that each user can keep track with the narrow beams at all times. This choice also leads to a demand for more base stations because the density of these has to be greater for effective communications.

Even at the end of the twentieth century the mobile satellite systems (MSS), the slow-uptake Iridium and the emergent Globalstar, remained essentially narrowband and did not enable dramatic change in terms of much more rapid Internet on-ramping. Broadband satellite and terrestrial systems, mainly using Ka-band, will be entering service early in the twenty-first century, and later on systems utilizing V-band are planned. These approaches are considered in Chapters 8 and 9.

However, component technologies remain a major challenge from specification and price aspects.

2.3 Bandwidth and frequency implications for component technologies

In the electronics world of the third millennium everyone will understandably want all of the component technology to be solid state, that is, semiconductor based. However, at the end of the twentieth century millimeter-wave power sources for transmitters were often tubes rather than semiconductors because only tubes can deliver the power levels required.

As a general rule, one that applies remarkably well to semiconductors, as frequency rises so the available output power falls. The faster you try to make them run the hotter they get, and so you try making them larger, but then they will not run so fast anyway. How many athletes, especially sprinters, weigh in at 300 pounds (134 kg) or more? Having said this, millimeter-wave semiconductor component technology continues making great strides and some turn of the millennium product capabilities for millimeter-wave MMICs are indicated in Figure 2.7.

Hewlett-Packard's and TriQuint's MMICs deliver continuous (CW) output power levels of 0.3–0.5W over the 36–40-GHz band. Quinstar manufactures Impatt chips providing CW power outputs of 0.25W at Q-band. There are many other companies in this field and the list grows rapidly.

At lower frequencies, through the subgigahertz cellular bands and up to around 10 GHz, CW output power levels from 1W to 3W are regularly obtained with MMICs manufactured by the likes of Alpha Industries and M/A-COM (owned by AMP, Inc.). The overall interesting point about Figure 2.7 is that a broad curve traced through the points indicates a continuous decline with frequency. Discrete semiconductors, as opposed to the MMICs applying above, provide more output power and can achieve 3W at 30 GHz. But these devices need substantial external circuitry to provide the desired system functions and power is lost by dissipation in the circuit elements.

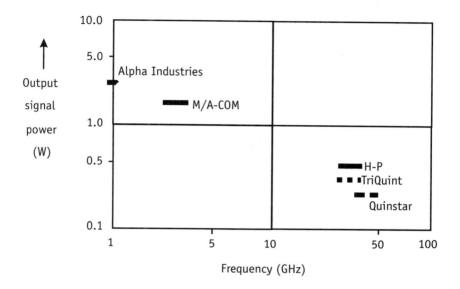

Figure 2.7 Power versus frequency for a selection of MMICs.

The use of multiple MMICs in power-combining configurations is almost certainly the way forward and there is a thrust of effort in this direction. Two techniques compete for this power combining: circuit-based (called "corporate") and spatial. In corporate power combining the outputs from many MMICs are fed into one single common output port. As the number of MMICs reaches into the high tens or hundreds range, so efficiency drops considerably and the losses soon exceed any gains due to the combining. This applies to combining levels definitely required for millimeter-wave because of the initial low power MMICs and the increased losses due to the high frequencies.

Spatial combining, where power from the individual chips is added by "funneling" the energy into a common waveguide, avoids the problem with the corporate losses. The main drawback is the relatively large dimensions with this technology, but this is less of a problem at higher frequencies because waveguide sizes then fall anyway. For example, the University of Santa Barbara (CA-MURI effort) has demonstrated spatial combiners in which a 20-GHz input signal is multiplied upwards to produce a 60-GHz output.

The above considerations are particularly significant for the realization of several types of systems described later, notably in Chapters 4, 6, 8, and 9 on defense, CATV, stratospheric and terrestrial, and satellite systems.

2.4 Radio and optical bandwidths

As mentioned earlier, the available bandwidth in a wireless link (also a fiber cable) depends critically upon the frequency (or wavelength) regime in which the channel is desired to operate.

Radio signal attenuation in the atmosphere is shown as a function of frequency in Figure 2.8. Air becomes less dense as altitude increases and therefore a signal to or from an antenna pointing vertically, elevation 90 (i.e., zenith), suffers least attenuation. In contrast, when the antenna is pointing in an almost horizontal direction the signal has to pass through much denser air and the attenuation is greatest. In summary, the total amount of atmospheric attenuation depends therefore upon the angle of elevation of the antenna.

A close study of Figure 2.8 reveals that the attenuation varies greatly with frequency, exhibiting highly pronounced peaks and troughs when the frequency exceeds 10 GHz. This behavior is typical of many material media and the peaks of attenuation are the result of oxygen and water vapor absorption. It can be seen that the main peaks occur at 22 GHz and 60 GHz. There are other peaks at 118 GHz, 183 GHz, and 325 GHz, but these extremely high frequencies are of little significance in telecommunications.

Some defense systems are deliberately tuned to frequencies such as 60 GHz because then allied signals can be locally "buried" by the attenuation. Local allied theater operations will accept the signals but more remote attempts by a distant eavesdropping enemy will be thwarted because of greatly attenuated received signals. Traditionally this has worked to some extent, but sensitive receiving techniques, especially when the signals are digital, enable very weak signals to be extracted from the burying noise background.

Most commercial systems are designed to operate within the frequency bands where relatively low attenuation applies. The first such

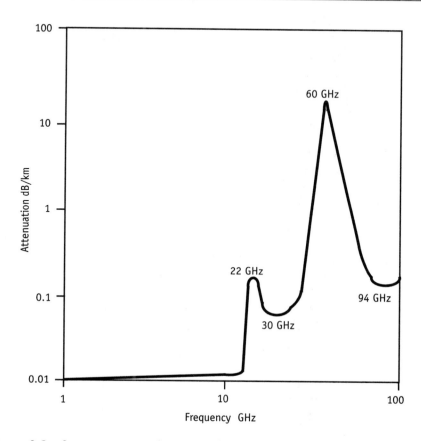

Figure 2.8   One-way attenuation versus frequency for microwave signals through the atmosphere.

band extends from approximately 25 GHz to 40 GHz and many systems utilize portions of this band. The next low attenuation band starts at about 77 GHz and stops around 105 GHz. In the center of this band an attenuation minimum occurs at 94 GHz and this frequency is used for some millimeter-wave radars. Because of component availability and costs the lower band-edge frequency, 77 GHz, is also popular but this is mainly for automotive cruise control radars.

It is by no means always necessary for systems to operate in the bands of lowest attenuation, but it is important to select frequency bands that are little used by existing terrestrial or satellite systems and facilitate reasonably "technologically ready" components. This applies to the Ka-band (26.5–40 GHz) and both Microsoft and McCaw Communications plan that this band will be used for the Teledesic system, for example. The very important LMDS networks also absorb portions of the Ka-band. However, other planned satellite systems such as Skybridge are expected to still use the crowded Ku-band (12–18 GHz), overcoming the crowded spectrum problem by spectrum-sharing. This is discussed in Chapter 9.

Later still in the twenty-first century, probably around 2009, V-band satellites are planned and these will have bearers located in the 46–56-GHz range. This band is halfway up the curve (see Figure 2.8) toward the 60-GHz peak and provides a 10-GHz bandwidth.

Precipitation, especially rain, also seriously affects signal transmission through the earth's atmosphere. Heavy rain, falling at 8 mm per hour, can add as much as 8 dB of extra attenuation to the signal at 30 GHz and 12 dB at 60 GHz.

As with the attenuation experienced by microwave and millimeter-wave signals traveling through earth's atmosphere, so also optical signals suffer highly nonlinear attenuation in glass (silica) fibers. Instead of frequency, the optical industry talks in terms of wavelengths, and the attenuation versus wavelength behavior for silica fiber is shown in Figure 2.9.

The general trend downwards in attenuation, at least up to 1,550 nm, is caused by decreasing scattering of the signal (known as Rayleigh scattering) and the two dips in the curve, centered on 1,320 nm and 1,552.5 nm respectively, are termed *transmission windows*. Unavoidable impurities in the glass system, particularly water in the form of O-H ions and silica in the nature of Si-O ions, cause this nonlinear attenuation. There is an O-H absorption peak around 1,370–1,380 nm and infrared absorption resulting from Si-O ions increases markedly above 1,660 nm. Had frequency again been used as the parameter instead of wavelength, the behavior would be seen as remarkably similar to that applying to radio waves through the atmosphere, that is, the trend of attenuation would always be generally upwards.

There is another potentially serious problem with optical fibers, namely dispersion. The main consequence of dispersion in any transmission medium is to broaden the digital pulses as these travel along the

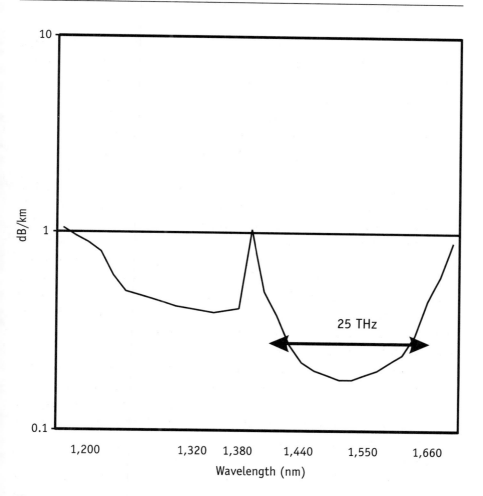

Figure 2.9   Single-mode optical fiber attenuation versus wavelength.

cable. If present to any significant extent this causes errors in the received bits so the bit error rate (BER) rises and reception is seriously impaired.

The first "window," around 1,320 nm, was originally the preferred center wavelength because fibers naturally offer zero dispersion of the signal pulses at this wavelength. Historically there was a conflict: either operate fiber at 1,320 nm and have the great advantage of zero dispersion, or operate at 1,552.5 (generally simply termed *1,550 nm*) where attenuation is lowest and also bandwidth greatest. Eventually the material choices in the fiber structure were designed so that the zero-dispersion wavelength coincided with the attenuation minimum of 1,550 nm.

This behavior applies to single-mode optical fiber which is used in all long distance and increasingly in very high-speed shorter links such as gigabit Ethernet. From Figure 2.9 it is easy to see why the 1,440-nm to 1,660-nm band, centered on 1,552.5 nm (193.1 THz), is exclusively used for such broadband systems. Translated into frequencies rather than wavelengths, the available bandwidth in this window amounts to no less than a staggering 25 THz (yes, that really is 25 THz, i.e., 25 million GHz). In terms of channel capacities or bit rate possibilities this is well out to the right in Figure 2.4. Terabits per second (Tbps) are now a real possibility and such rates were achieved for practical systems in the latter years of the twentieth century. This bandwidth is 2,500 times the 10-GHz millimeter-wave bandwidth referred to earlier. The maximum bit rate supportable by the 10-GHz band could theoretically be over 2,500 times larger with the optical channel!

Erbium-doped fiber amplifiers (EDFAs) are also available for lightwave signal amplification within this band although none can cover more than about one-eighth of the entire band. Such amplifiers typically provide around 20 dB to 25 dB of gain over 3-THz bandwidths, and a typical example of such an amplifier is shown in Figure 2.10. Fujitsu and Hewlett-Packard are examples of corporations that manufacture these types of products, the basic arrangement of which is shown in Figure 2.11.

As always with any amplifier there is a signal input port and a signal output port. The source of optical power, from which the signal gain is derived, is a semiconductor laser that usually operates at 980 nm and the output from this laser feeds into a wavelength division multiplexer (WDM). The combined input signal and the "WDM'd" power from the laser are fed through a length of erbium-doped fiber in which the amplification actually takes place. An output optical isolator ensures that negligible amplified and reflected output power reenters the amplifier (this could cause oscillations—or worse). WDM is described in Chapter 3.

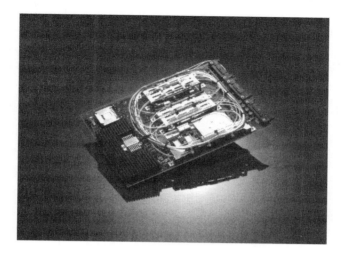

Figure 2.10 Broadband optical amplifier. (Courtesy of Fujitsu.)

EDFAs are now vital elements in long-distance optical systems and are becoming increasingly important for very-high-bit rate moderate-range networks.

2.5 Cost implications as bandwidth increases

As a very general rule with plenty of exceptions, for radio systems costs tend to increase as bandwidth demands increase. Historically this has proved to be a critical problem, and the costs associated with direct-broadcast by satellite (DBS) have been kept down to consumer-affordable levels by virtue of a narrow bandwidth (in relative terms) and of course with volume manufacturing. It is always the advent of volume manufacturing driven by steadily expanding demand that enables the prices of systems components to fall. Witness the trend with PCs, DBS, and even "quality" cars over the past years. Initially such products are affordable by the comparatively wealthy few, but as volume demand increases unit prices fall, market volumes increase, and suppliers' operating margins tend to become squeezed.

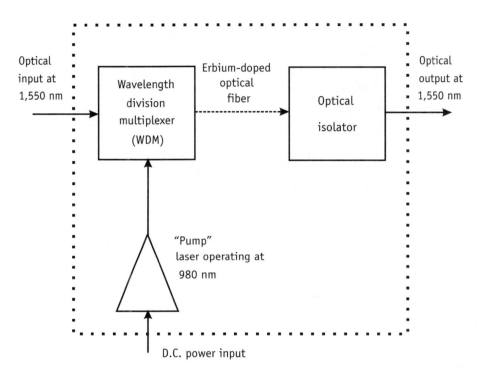

Figure 2.11   Functional schematic of the EDFA.

In contrast, traditionally the military often requires relatively wide bandwidths, a prime example being electronic warfare (EW) systems. While cost is today a much more significant parameter for the defense industry than it was back in the cold war days, this cannot change the fundamental requirement for broadband EW systems and "octave" bandwidths are commonly encountered. By *octave* we always mean a two-fold frequency range, for example, from 8 GHz to 16 GHz, or from 13 GHz to 26 GHz.

In commercial systems, particularly when high volumes are demanded, marginal costs are critical and any means for minimizing bandwidths are taken up enthusiastically.

Taking the V-band defined above the entire bandwidth is 10 GHz (i.e., $56 - 46$ GHz) but it is only possible to utilize fractions of this wide range for reasons associated with economics and systems planning.

Several types of systems must be crowded into this wide frequency span and mutual interference must be minimized. Also, systems components such as amplifiers covering the entire 10-GHz band would be prohibitively expensive for end-users—at least until volume manufacturing happens.

As broadband systems steadily accelerate in implementation, volumes will increase, substantially for many final end-user requirements, and prices will fall as a consequence of these increasing volumes. Year 2015 millimeter-wave module unit prices will equate to the price levels previously associated with late-twentieth-century Ku-band modules. Recent examples of this trend abound. In 1995 typical unit prices for MMIC power amplifiers, operating over low microwave frequencies, were around $200 in 100–1,000 volumes. By the end of the second millennium these prices had dropped by a factor of ten. By contrast, the 1998 unit prices of more advanced, faster, MMICs remained in the $300 (e.g., 10-Gbps amplifiers) to $700 (14–14.5-GHz 20-dBm amplifiers) range. With increasing volumes these prices will undoubtedly fall during the 2000-plus period.

Interestingly with optical systems the cost pattern does not follow that of the radio scenario.

Costs are actually lowest for the multimode fiber optics used in applications such as CATV, and this is due to a combination of larger-size components (the cables, connectors, and optoelectronics) and also volume deployment. Yet it is with these types of systems that the shortest wavelengths are used, that is, the highest optical frequencies.

As the frequency decreases (or, correspondingly, the wavelength increases), so that the wide third window centered at 1,550 nm is available, production costs and unit prices actually increase. This is because single-mode fiber must now be used with the attendant much more critical and small physical dimensions. However, one common factor is that bandwidth has dramatically increased.

So, the behavior with regard to fiber optics (lightwave/photonics) costs follows an opposite trend in terms of frequency from that experienced with regard to radio systems.

The moral is to beware of extending conventional wisdom from one type of technology available in any one era to another contrasting technology fit for a different time and different applications.

Select bibliography

Balston, D.M., and R.C.V. Macario, *Cellular Radio Systems*, Norwood, MA: Artech House, Inc., 1994.

Brodsky, I., *Wireless: The Revolution in Personal Telecommunications*, Norwood, MA: Artech House, Inc., 1995.

Elbert, B., *Introduction to Satellite Communication, Second Edition*, Norwood, MA: Artech House, Inc., 1999.

"Forward and Derivatives Trading of Telecoms Capacity," Risk Conferences held at the Churchill InterContinental Hotel, London, U.K., June 14–15, 1999.

Freeman, R.L., *Telecommunication System Engineering*, 3rd ed., New York: John Wiley & Sons, Inc., 1996.

Fujitsu product information on the EDFA at www.fujitsu.co.jp/index-e.html

Microwaves & RF, Several articles in the Communications Issue, April 1999.

Nellist, J.G., and E.M. Gilbert, *Understanding Modern Telecommunications and the Information Superhighway*, Norwood, MA: Artech House, Inc., 1999.

Telecommunications, Editorial item on bandwidth trading, June 1998, p. 20.

Willner, A.E., "Mining the Optical Bandwidth for a Terabit per Second," *IEEE Spectrum*, April 1997, pp. 32–41.

3

Broadband Cabled Networks

3.1 Multimode or single-mode fibers?

Until the final two decades of the twentieth century *cabled* always meant a metallic conducting electrical connection between several points. The metal cables were predominantly of copper for both power and telecommunications. In the case of telecommunications, the basic choice was between twisted pairs, coaxial transmission lines, or metallic trunk pipes known as waveguides.

By the mid-1960s it was recognized that bandwidth demands would soon be outstripping the capabilities of either twisted pairs or coaxial cables, and the pressure was on to do something about the imminent problem. Several research organizations, notably in Atlanta (Georgia, U.S.), New Jersey (U.S.), Japan, and the United Kingdom, began work based upon the use of circular metallic waveguides for terrestrial broadband transmission. While in purely technical terms this appeared to provide a most promising solution, the practical and economic implications were highly problematic.

A specific mode in circular waveguide, known as the TE_{01} mode, is unique in that it offers attenuation amounting to less than 3 dB per km over the entire frequency range from 30 GHz to almost 100 GHz. This was—and still is—an extremely broad frequency range by any standards for electrical signals. For such waveguides in straight runs above ground level the attenuation is less than 1 dB per kilometer from 60 GHz to 90 GHz. However, when the waveguides were buried below ground the attenuation doubled or even tripled, and in any case the slightest bending, deflections, or inconsistencies effectively ruined the critical performance.

Although much excitement was generated in the 1970s concerning this technology, it was soon appreciated that these practical difficulties and the economics of the operation would conspire to prevent the commercial implementation of the approach.

Meanwhile, in 1966, a research group headed by two now-famous researchers named Kao and Hockham, working at what is today Nortel's U.K. research center, published a landmark theoretical paper on the transmission of optical signals in glass strands. The paper predicted that such signals could be successfully transmitted along silica or glass "waveguides," but there was one somewhat serious problem: the attenuation would probably be in the region of some tens of decibels per kilometer. At the time this provided further fuel for the copper circular waveguide enthusiasts with their "extremely low" 3 dB per km.

However, by 1970 Kapron and Keck (U.S.) had achieved losses of "only" 20 dB per km in optical fibers, and in 1976 the Rediffusion Company (London, U.K.) introduced an optical fiber cable connecting cable TV subscribers as the first commercial application. In 1982 the first single-mode fiber cables emerged with losses down to an incredible 0.2 dB per km—an order of magnitude below the best copper waveguide results. As indicated in Chapter 2, this figure is very close to the absolutely lowest limit achievable for such cables.

Using conventional copper cables, even twisted pairs for very short tens of meters distances, bit rates exceeding 30 Mbps are achievable without any recourse to digital subscriber line (xDSL) techniques. The adoption of xDSL approaches means that higher bit rates can be supported over longer reach distances. For example, with ADSL 6M a bit rate of 6 Mbps can be transmitted over a distance of 2 miles (see Figure 3.1). Lower bit rates can be transmitted over distances up to at least five miles (8 km) using what is sometimes termed *copper optics,* that is, fiber modulation schemes extended to copper transmission lines.

Worldwide the markets for fiber-optic equipment of all forms—cables, connectors, transceivers, amplifiers, test equipment, and so forth—continue to grow steadily and this is expected to continue and

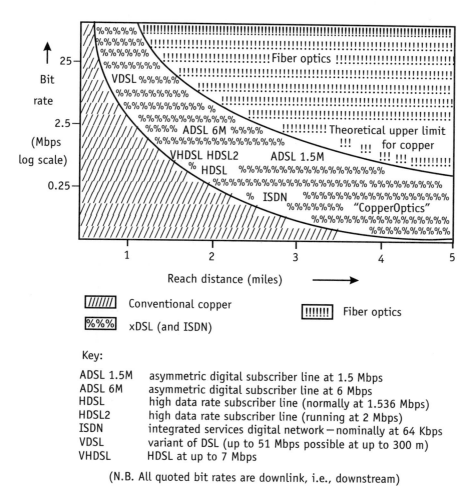

Figure 3.1 xDSL bit rates and reach distances.

indeed accelerate well into the third millennium. The chart shown in Figure 3.2 (consolidated from several sources) indicates this global growth and tens of billions of U.S. dollar market values are anticipated by 2004.

In Chapter 1 the concepts and some applications of fiber optics are described. There it is pointed out that most CATV networks of the late-1990s used fiber transmission but that most were what is termed *multimode*.

To the uninitiated it may at first seem that multimode cabling must surely be superior technically to single-mode, but in fact the opposite is the case. In multimode optical fibers many modes or "rays" of transmission are set up and these travel down the fibers in zigzag fashion. Also the optimum wavelength for transmission is typically 630 nm, 820 nm, or 1,300 nm, and fibers exhibit considerably more attenuation at these wavelengths than at the 1,550-nm optimum applying to single-mode. At best we are talking several decibels compared with 0.2 dB. It is not by any means only this attenuation feature that leads to the reach distance limitations with multimode fibers, because another important restriction known as dispersion represents a serious problem.

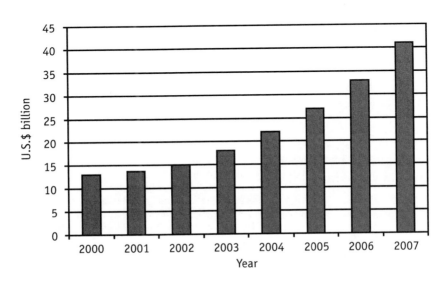

Figure 3.2 Global markets for fiber-optic equipment.

All fibers of all types, and in fact all transmission media, tend naturally to exhibit dispersion. The zigzag trajectories referred to above are a major source of dispersion in multimode fibers and this is characterized by different times-of-arrival for different rays as these travel along the length of cable. As a result digital signals arrive out-of-sync at the receiver—some are delayed compared with others. Beyond a definite limit this dispersion produces what is termed intersymbol interference (ISI) at a level which is intolerable for the system's operation. Controlling this ISI is a major goal for the system's designers.

There is, however, a highly significant advantage with multimode fiber cables and this is the relative ease and economy of interconnections, including the demountable components. This applies because the dimensions of the multimode cables are inherently much larger than those for single-mode: tens-to-hundreds of microns compared with several microns.

A typical multimode fiber structure will have a 50- or 62.5-micron core (the smallest internal diameter) and a 125- or 150-micron outer diameter. In comparison, a single-mode fiber has an 8-micron core and a 50-micron outer dimension. Handling the multimode cable is much easier and more reliable than is the case with single-mode.

This fact continues to place emphasis upon multimode technology wherever possible, and the "power penalty" for maintaining ISI at tolerable levels is shown in Figure 3.3, where the parameter is the fiber bandwidth × distance product (often simply termed *bandwidth*) in megahertz per kilometer. The power penalty is a measure of the amount of additional transmitter power required to maintain the tolerable ISI.

At 500 MHz per km the best fiber shown in Figure 3.3 is essentially state-of-the-art for the turn of the millennium, and even over half a kilometer reach distance, less than 3 dB power penalty is incurred.

In summary, multimode fibers are mainly used in applications such as late twentieth-century CATV, LANs, and the local loop.

Single-mode fibers are principally used in long-distance, often suboceanic, national, and international trunk cabled applications and also in some advanced shorter routes. Nellist and Gilbert's book, *Understanding Modern Telecommunications and the Information Superhighway*, covers these types of applications well. A further major advance will almost certainly come from what is termed *soliton transmission* in which special pulse shapes

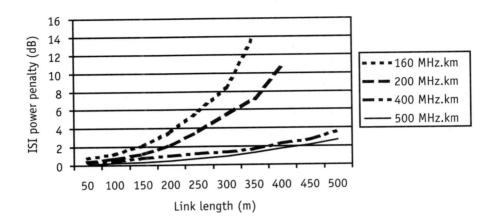

Figure 3.3   ISI penalties for different fiber bandwidths.

are transmitted enabling tens of kilometers to be spanned without intervening amplification and regeneration (Nortel and others).

In this book the main focus is on shorter distance cabled networks, but SDH and SONET, introduced in Chapter 1, are of such importance that they are considered next.

3.2 SDH, SONET, and fiber

As mentioned in Chapter 1, digital signal framing and multiplexing are of major importance. This leads, in turn, to basic standard requirements for bit rates (or data rates) for transmission at various levels through the networks.

Historically the bit rates have differed in various countries and world regions as a result of earlier evolution that was based largely upon national considerations. Countries such as Japan, the United States, the United Kingdom, and European nations all had somewhat different and mutually incompatible digital hierarchies. In North America, for example, the digital signal (DS) hierarchy series was prevalent although this involved noninteger multiples of the basic PCM voice channel rate (64 Kbps) at the first two levels: DS-1 and DS-2. Figure 2.2 of Chapter 2 indicates the multiplexing scheme and here only the DS-3 level is an integer multiple of 64 Kbps ($\times$ 699 exactly). DS-3 is the same as T-3, DS-4 is the same as T-4, and so on.

In the latter portion of the twentieth century *globalism* became the watchword and it was clearly necessary for every nation and region to be able to intercommunicate without having to translate from one digital hierarchy to another different system. Apart from hardware and software requirements, this would also risk substantial timing errors in the digital signals. As a result the International Telecommunications Union (ITU) agreed upon what is now universally known as the synchronous digital hierarchy, or SDH. Apart from the first optical level, known as OC-1, all the SDH rates are precisely equal to corresponding optical rates as shown in Table 3.1. Timing errors are kept to a minimum by adopting atomic frequency standards and GPS.

In Table 3.1 OC means optical carrier, SONET refers to synchronous optical network, STM means synchronous transport module (i.e., the information frame structure), and STS means synchronous transport signal, which is a term used in conjunction with SONET. Although the primary focus here is upon optical networks, it is important to understand that all the SDH (and therefore also STM-N) references apply equally to microwave and millimeter-wave radio systems including satellite. Therefore reference will again be made to these SDH levels later in this book.

Table 3.1
SDH and SONET Digital Signal Bit Rates

ITU-T: SDH (STM-m)	SONET (STS-n)	Optical Carrier (OC-n)	Bit Rates: Mbps, Gbps	Basic PCM Rate (64 Kbps) Multiplier	Voice Equivalent Channels
	STS-1	OC-1	51.84 Mbps	810	672
STM-1	STS-3	OC-3	155.52 Mbps	2430	1344*
STM-4	STS-12	OC-12	622.08 Mbps	9720	5376
STM-16	STS-48	OC-48	2.48832 Gbps	38250	21504
STM-64	STS-192	OC-192	9.953 Gbps	153000	86016

* This is the "gateway" level, that is, the principal interconnection rate for connected networks.

Information for Table 3.1 was largely taken from Nellist and Gilbert's book referred to above (their Table 1.1), and this source is acknowledged.

The voice equivalent channels are calculated by dividing the bit rates by the basic PCM voice channel rate (64 Kbps) and allowing for various overheads applying to the STM and STS modules. This data is for general guidance purposes only since rarely if ever are the high-level channels used for voice exclusively, and in fact Internet data is increasingly taking the lion's share of the traffic.

The STM-m bit rates referred to in Table 3.1 increase progressively by factors of four. However, two intermediate rates may become introduced, namely STM-8 and STM-12—being 1.24416 Gbps and 1.866240 Gbps, respectively. Also, as bit rates continue increasing, at least two further levels will probably enter the sequence:

$$2 \times 9.953 \text{ Gbps} = 19.906 \text{ Gbps; and } 2 \times 19,906 \text{ Gbps} = 39.812 \text{ Gbps}$$

These would logically be termed STM-128 and STM-256.

A block diagram illustrating how the SDH signals are developed is provided as Figure 3.4. The aim is to group several signals to form an STM-16 output information stream at 2.488 Gbps. A wide variety of input signals is shown, ranging from the D1 (or DS-1) 1.544-Mbps signal, through the 2.048-Mbps European E1 level, followed by the 6.312 DS-2 and 44.736-Mbps DS-3 instances, and finally the high bit rate DS-4 at 139.264 Mbps.

The components C-n are data containers, AN are aligning units, MUX are digital multiplexers, and AUG is the so-called administrative unit group feeding the final output. Note how the lowest levels (i.e., slowest) input signals require the greatest amount of aligning and multiplexing before reaching the final output multiplexer. In contrast the fastest input signal (139.264 Mbps) can be passed directly into the final MUX.

The technology used here is generally conventional silicon electronics, although at the higher bit rates (notably at or above STM-64) silicon is insufficiently fast and the ICs need to be gallium arsenide (GaAs) or some other appropriate semiconductor.

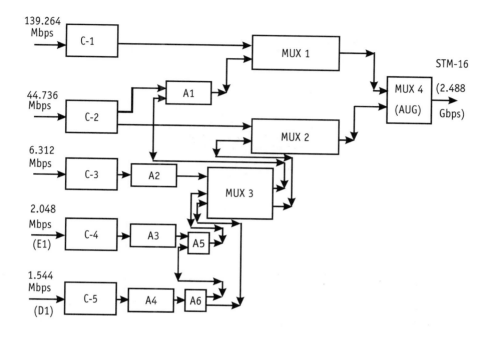

Figure 3.4 Examples of the SDH multiplexing structure.

Where the output signals are transmitted using fiber optics specialized components must be used and many suppliers provide products meeting these requirements. As an example the photograph in Figure 3.5 shows such a hand-sized product manufactured by Hewlett-Packard. The HFCT-1027 operates over the 50 Mbps to 622 Mbps (STM-4) range, has a reach distance of 400 km at the upper limit, and uses an internal laser working at 1,530–1,565 nm. Products with broadly similar configurations operate with bit rates as high as 2.448 Gbps, that is, STM-16.

A global market forecast for SDH equipment is shown in Figure 3.6. It should be appreciated that the data here is for SDH applied to all types of transmission (not only fiber optics).

Although these "billion dollar" markets increase until the year 2001, after this time a steady decline sets in resulting from a combination of market saturation and inroads from new technologies such as dense wavelength division multiplexing (DWDM is covered in Section 3.4).

Figure 3.5 50–622-Mbps (STM-4) transmitter for single-mode fiber. (Courtesy of Hewlett-Packard.)

3.3 Gigabit data network evolution

Local area networks (LANs) have established a secure niche in data communications. Although wireless LANs (WLANs) are now of growing importance, these have a fair way to go before broadband capabilities become available and most LANs remain predominantly cabled.

Early LANs supported backbone (bus) bit rates only as high as about 10 Mbps, but fiber-based networks soon pushed this up to 100 Mbps—with the fiber distributed data interface (FDDI) and fast Ethernet.

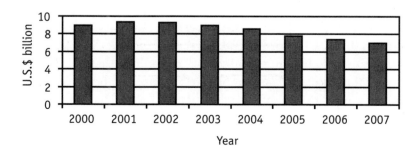

Figure 3.6 Global SDH equipment market forecast.

FDDI does not have to use fiber as the bus physical layer, since the name refers to a specific transmission protocol. Where link lengths are sufficiently short (generally much less than 500m) and interference is not too much of a problem, copper conductors may be used.

Although not strictly a broadband data network by current standards, the FDDI LAN backbone is briefly considered here because it paved the way for much faster standards.

This LAN protocol normally employs 4B/5B coding that allows for a maximum line transmission rate of almost 125 Mbps. This 4B/5B coding operates by taking four bits of the signal in blocks and the approach is considerably more spectrally efficient than many other methods. Using fiber as the transmission medium, total path lengths up to 200 km are possible, that is, 100 km of single fiber because a dual fiber configuration is implemented. With the basic protocol 500 nodes or stations can be supported but parameter values may be increased to allow for much larger networks.

Simplified by identifying only six stations, an overall schematic block diagram of FDDI is shown in Figure 3.7. In this example each active node (station) regenerates the required signals. In practice many more nodes are implemented and one will be the server station. Some stations just operate in bypass mode, not playing a part in the LAN operation, and this is illustrated with station number 3 here.

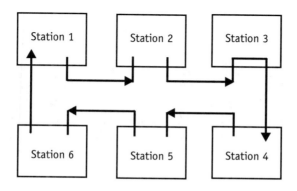

Figure 3.7   Schematic block diagram of FDDI.

FDDI basically operates on the token ring principle in which a unique control signal, known as the token, circulates on the transmission bus medium. Only the active stations are primed to react to the presence of the token which is captured and removed from the ring before that station transmits its message signal. At this time the station issues a new token.

Some stations can actually comprise complete "sub-LANs" in their own right.

For fiber-based FDDI LAN backbones, multimode cables are generally implemented, with wavelengths of either 820 nm or 1,300 nm.

As mentioned above, LANs began with exclusively copper transmission layers and many used the so-called Ethernet protocol that was originally developed by Xerox Corporation many years ago. The fundamental Ethernet topology is shown in Figure 3.8. Only six stations are shown connecting to the bus bidirectionally, and in practice these types of LANs cannot support large numbers of stations where copper transmission applies.

The bit rate standard is a low 10 Mbps for copper-based Ethernets and carrier-sense multiple access with collision detection (CSMA/CD) is the protocol. With this a given station only transmits when no data signal is detected as being present at the node—therefore avoiding collisions.

With copper-based Ethernets, standing waves can be a problem on the transmission bus and this imposes restrictions—not least on any attempts to speed up the electrical versions of these LANs. The "terminations" (matched loads) assist in partially solving this problem but limitations remain.

As unit prices for gigabit-level fiber-optic components continued to fall during the late twentieth century, so systems integrators seized the opportunity to introduce this technology into LANs to a much greater extent than ever before. The dramatic growth of all forms of communications—particularly data—provided a continuous driving force.

The simplicity of the Ethernet topology with its ready extendibility led to the development of what has become known as "Gigabit Ethernet," which means the basic Ethernet topology but implemented with a fiber transmission bus operating at 1 Gbps or higher.

One example of a state-of-the-art product in this area is Lucent Technologies' Cajun P550 Gigabit Switch and Routing Switch that has a backplane capacity of 45.76 Gbps. This product can support up to 33 million

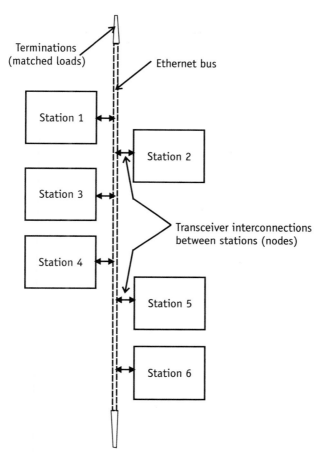

Figure 3.8 The basic Ethernet LAN topology.

packets per second (pps) of Layer 2 switching and up to 18 million pps routed into and out of the network via the routing switch. The Cajun P550 can interoperate with up to 1,000 virtual LANs (VLANs)—LANs formed by using various existing network elements that by themselves are not classed as individual LANs.

Some of the latest developments in fiber networks blend together the fundamental approaches and technologies associated with ATM, SONET, and LANs. This leads to the important concept of "points of presence" and such developments are introduced next.

3.4 Gigabit junction and GigaPOPs

With data communications, especially Internet-type traffic, continuing in a lengthy phase of explosive growth by the late 1990s, it was abundantly clear that something had to be done in order to relieve congestion and prevent gridlock. This was fast becoming a real problem in scenarios where increasingly vast files of information were being exchanged—campuses and their interlinking being a prime example.

Conventional LANs just would not cope with the new situation—not even Gigabit Ethernet.

Driven by this acceleration in demand the U.S. government announced its next generation Internet initiative and this was something that might be termed *Internet2*. Within the United States several regional networking organizations developed various approaches toward solutions for this problem. Among these organizations were the California Research and Education Network 2 (Calren2), the Metropolitan REN (MREN, in Chicago), the Houston Area Computational Science Consortium (HACSC), the New York State Educational and Research Network (Nysernet), and the North Carolina Networking Initiative (NCNI).

The material presented here is based upon NCNI's gigabit point of presence, or GigaPOP (or gigabit junction), although in this presentation the topologies could represent any subsequent development with similar basic requirements.

The overall aim was to provide a high degree of flexibility and adaptability within a network sufficiently advanced and broadband to enable heavy data users to communicate efficiently without bottlenecks. It is also vital to configure the network using standard and commercially available components in an economical topology. These aims are mainly met with the GigaPOP as shown in Figure 3.9. This supports three major user networks on its main network (Nm), and in the case of NCNI the network vendors (V1, V2, and V3) are Cisco, IBM, and Nortel, all located at Research Triangle Park. New networks can readily be added.

Four other major networks are also supported, N1 to N4 in Figure 3.9, and these would typically be at a range of different campuses. Again, from any of these networks further subnetworks can be added and one is shown as "New Network 2," connected via N1 using an Internet Protocol (IP) router.

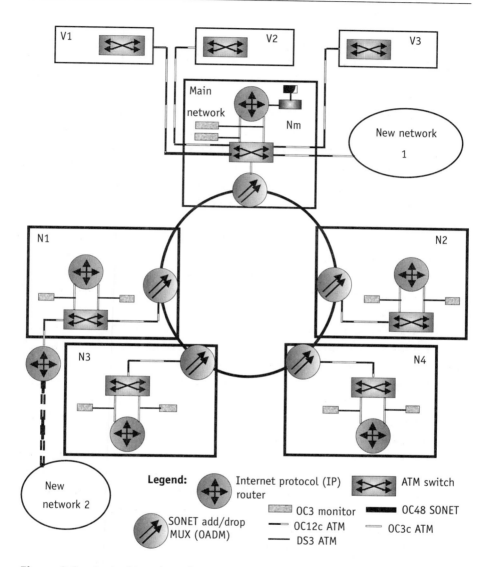

Figure 3.9 A gigabit point-of-presence network (GigaPOP).

The central core of the overall network is an OC-48 SONET single-mode fiber ring running ATM signals and operating at 2.488 Gbps. The SONET ring has optical add/drop multiplexers (OADM) located at a total of five nodes around it and these provide very high-speed duplex interconnections for each attached network. From each OADM (supplied by Nortel for the NCNI GigaPOP) the next stage is an ATM switch that provides the switching fabric for the routers and network monitoring.

Connections between OADMs and ATM switches are all via further single-mode fiber cables running OC-12c ATM at 622.08 Mbps. This also applies to subsequent connections from ATM switches to other ATM switches or into routers serving large new networks. These are all shown as semifilled interconnections in Figure 3.9.

IP routers are all fed from ATM switches by dual multimode fiber cabling operating on OC-3c at 155.52 Mbps, and each cable connection is monitored to ensure that network status and stability are maintained. Monitors are present on both fiber connections between switches and routers in all networks. Additionally, the main network also feeds a more comprehensive surveying subsystem that monitors and manages the entire GigaPOP.

For the NCNI network, Cisco Systems supplied the ATM switch-to-router links running at 155.52 Mbps. These OC-3c connections are all illustrated as "clear" (unfilled) cables in Figure 3.9.

Connections between the monitors and the ATM switch-to-router links are at the North American DS-3 ATM standard rate of 44.736 Mbps. This bit rate as well as the relatively short distances involved allows copper cabling to be implemented.

Not shown in Figure 3.9, for simplicity, many networks of varying complexity are connected to the IP routers. These range from highly localized simple LANs such as 10-Mbps Ethernets or token rings, operating at bit rates in the low Mbps to DS-3 or somewhat higher, up to sub-networks like FDDI or even Gigabit Ethernet. Attached to these LANs will be hundreds or thousands of PC terminals, central printers, and servers.

SONET line cards are relatively expensive units, certainly when compared with ATM switches, and an optimum network topology can be chosen that greatly reduces the number of SONET cards needed overall. In the topology of Figure 3.10(a) a total of 16 line cards are required—accounting for one card per input or output port on each traffic route into and out of each OADM. However, by outputting through an ATM switch unit at every OADM and routing the signal through these

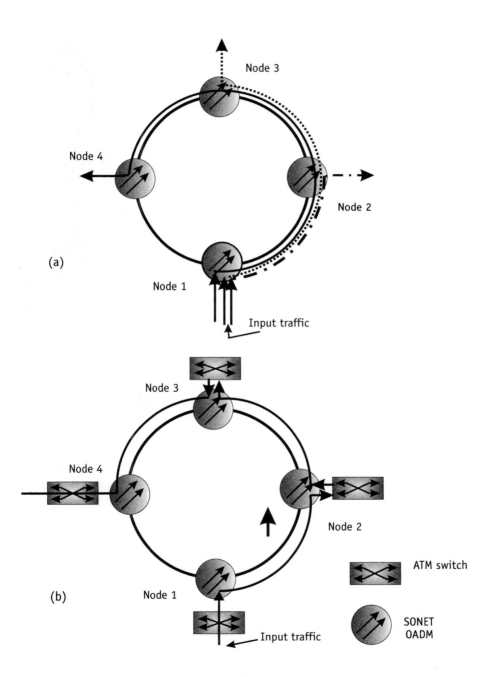

Figure 3.10 Using ATM switches to reduce costs.

units [Figure 3.10(b)] only seven SONET line cards are needed. With this type of approach costs are reduced considerably.

In summary, although as always solving problems in advanced high-speed networking is challenging, the ever-increasing availability of many types of sophisticated products enables new topologies to be implemented that otherwise could never be seriously considered. Cost reducing approaches should also always be followed.

One of several future advances under study for the enhancement of GigaPOPs is the use of DWDM. This technology has entered the trunk telecommunications fiber optics scene very strongly and it is described in the next section.

3.5 From WDM to DWDM

Frequency-division multiplexing (FDM), applicable to electrical signals, is described in Chapter 2, where Figure 2.1 illustrates the principle.

In the optical world, wavelengths are conventionally referred to rather than frequencies but the basic principle is the same, that is, information channels are segregated into different wavelength ranges. By analogy with the electrical case this process is termed wavelength-division multiplexing (WDM). Within each wavelength range the optical signals suffer a small amount of attenuation whereas beyond the allocated channels the attenuation rapidly rises to typically several tens of decibels.

WDM is mainly applied to single-mode systems but it can also be used with multimode networks. The principle is shown in Figure 3.11. Light signals from laser sources operating at different wavelengths are fed in parallel into the wavelength multiplexer which then outputs a single continuous optical signal made up from all "N" of the input wavelengths. The fiber cable carrying this combined signal must have sufficient bandwidth to be able to cope with what is often a very large capacity. However, the bandwidth of single-mode fiber at 1,550 nm is extremely large (see Figure 2.9 in Chapter 2).

At or close to the receiver a passive star coupler takes this combined signal as its input and outputs an attenuated version into "N" output ports, where N is the number of original wavelengths. Following each output port a tunable filter selects one of the required wavelengths (rejecting all

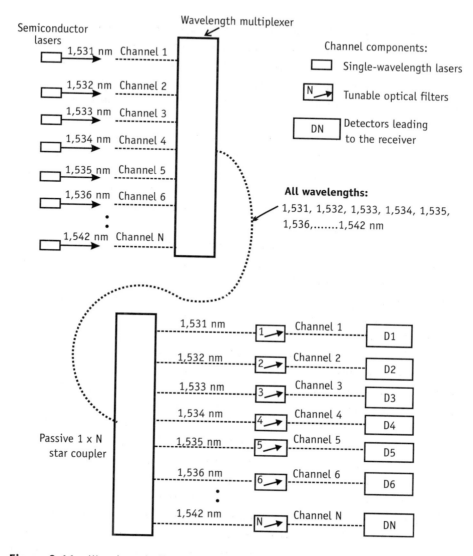

Figure 3.11 Wavelength division multiplexing.

others) and passes this signal to the receiving detector. The final signal is subsequently delivered to the user after conversion to electrical form.

A passive star coupler is an optical component with one input port and at least as many output ports as there are wavelengths. Wavelength tuning filters are made available by exploiting the wavelength-selective properties of passive optical devices such as diffraction gratings, electro-optic filters, and acoustic-optic filters. The actual tuning can be achieved mechanically, for example by rotating a diffraction grating, or electronically by attaching the filter to a semiconductor chip and varying the applied voltage.

Most WDM components function using miniaturized optical structures based upon diffraction grating or Bragg principles in which light beams emitted by the different lasers become focused into a single stream that is coupled into the output fiber. All WDM components are therefore passive because no semiconductor or other oscillating, amplifying, or switching function is involved.

WDM first became available during the 1980s when Deutsche Telecom, for example, introduced the technology into their BIGFON metropolitan networks. Ten-channel WDMs were implemented with 36-nm channel spacings and 20-nm signal pass bandwidths. Manufacturers included AEG (now a part of Daimler-Benz), Amphenol, and Corning-France.

By the turn of the millennium WDM technology had not only become much more sophisticated—it had also become vastly denser and had entered major growth markets. Late twentieth century WDM was mainly being termed *dense-wavelength division multiplexing* (DWDM). With this technology the channel count raced towards the 130-plus realm—unthinkable by 1980s standards. In 1999, for example, Lucent Technologies won its first order for an 80-channel system, and Pirelli offered a system that was scalable to 128 channels.

For most systems integrators it is the implementation of some degree of "future-proofing" that is so significant because the economics of repurchasing and reinstalling entirely new systems places substantial financial burdens on the companies. Anything—any technology—that is affordable now and yet offers the real prospect of being inherently upgraded in the future is likely to win important votes at the boardroom level.

While most 1999–2000 systems operators were typically only using six to eight of their available 80 to 128 channels, by year 2005 or so many more are expected to enter service.

DWDM manufacturers include Ciena, Fujitsu Telecoms, Lucent, and Pirelli (who also supply integrated subsystems). Ciena/Juniper, for example, offers an OC-48 router with a capacity of 40 Gbps transmitting over a distance of 12,000 km. Lucent Technologies offers their WaveStar Optical Line System (OLS) 400G. A modular optical transmission system, WaveStar OLS 400G is an 80-channel optical line system (OLS) for global networks that delivers up to 400-Gbps capacity over a single strand of fiber. This is equivalent to carrying more than three million simultaneous calls on one fiber.

Pirelli supplies full-scale modular units embodying DWDM, some of which are bidirectional. A front view of Pirelli's "Multi-Wave" product is shown in Figure 3.12 where the complexity is evident.

Philips Broadband Networks (Atlanta, GA) offers optical transmission systems incorporating DWDM up to 128 channels (digital supertrunking), and Ericsson Business Networks (Stockholm, Sweden) has a WDM platform that they term ERION.

Figure 3.12 A multiwavelength system. (T31 courtesy of Pirelli.)

Several vendors are also offering DWDM on 1,300 nm-based multi-mode fiber networks with a focus upon cable TV applications.

WDM and DWDM technologies complement TDM (described in Chapter 2), and signals have to be interleaved in order to optimize the capacity of optical fiber networks. In other words both DWDM and TDM must be combined, with the TDM "envelope" containing the DWDM optical signal. The effects of this approach are clear when comparing channel capacities and this is shown in Figure 3.13.

Transmission capacities are always maximized when both DWDM (or WDM) and TDM are used together, whereas with TDM or ATM alone the bit rates are always lower by around two orders of magnitude.

Global markets for DWDM equipment, indicated by the bar chart of Figure 3.14, are growing strongly and should approach U.S. $10 billion by the year 2007. This data is based upon public domain KMI-advertised data, moderated somewhat to reflect Y2K effects and extended to the year 2007. In Figure 3.14 LD means long-distance, SD means short-distance, and the cross-hatched bars refer to the retrofit market. All market segments are continuously growing for this exciting technology.

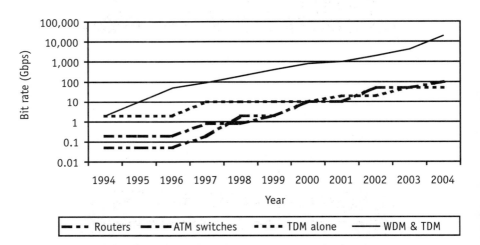

Figure 3.13 Bit rates for legacy and advanced multiplexing fabrics. (Courtesy of Nortel Networks.)

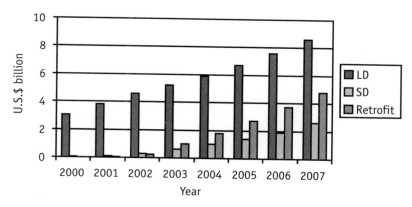

Figure 3.14 Global markets for DWDM.

3.6 Summary

Fiber must be used in broadband cabled links in the following instances:

- Reach distances up to a few kilometers and bit rates exceeding 30 Mbps;

- Reach distances up to 4 or 5 km and bit rates exceeding 9 Mbps;

- Reach distances around and greater than 8 km and bit rates exceeding 4 Mbps;

- Almost any high-security cabled system.

Since the 1966 to 1970 period (Kao and Hockham, and Kapron and Keck) dramatic strides have been made concerning the capabilities of both multimode and single-mode fiber cables. Most CATV companies currently use the narrower-band multimode cables in their networks, but broadband single-mode upgrades are almost inevitable in the future. Long-distance trunk cabled communications have exploited single-mode for many years.

Ever-larger file transfer sizes, coupled with timing that makes sense to users, increasingly demand Gbps LAN connectivity and this has led to the Gigabit Ethernet. GigaPOPs require OC-48 SONET at 2.488 Gbps using ATM transmission and switches. All of these systems make extensive use of fiber cables.

DWDM is increasingly employed in high bandwidth networks where current markets are already huge and future growth is extremely healthy. Fiber amplifiers are currently mainly EDFAs, but intrinsically fiber-based amplifiers are probably ripe for exploitation in the near future.

Select bibliography

"Broadband Special," *Telecom Product News* (TPN), Supplement, April 1999.

Broadhead, S., "The Slow March of DSL," *Telecommunications*, March 1999, pp. 31–36.

Collins, J.C., et al. "Data Express: A Gigabit Junction with the Next Generation Internet," *IEEE Spectrum*, February 1999, pp.18–25.

Flanigan, B., "Fibre-rich Futures: The SDH Story," *Telecommunications*, February 1998, pp. 37–40.

Gannon, P., and A. Stewart, "Europe Sees the Light," *Communications International*, July 1998, pp. 9–14.

Gibson, K., "Towards a Photonic Future," *Telecommunications*, March 1999, pp. 63–72.

Hewlett-Packard (H-P) fiber-optic components, product selection information 1999, at http://www.hp.com/go.fiber.

Kobayashi, K.W., et al., "InP-based HBT Technology for Next-generation Lightwave Communications," *Microwave Journal*, June 1998, pp. 22–38.

Lange, L., "The Internet (Technology 1999 Analysis & Forecast)," *IEEE Spectrum*, January 1999, pp. 35–40.

Lombardi, P.-J., "Solving the Gigabit Challenge," *Telecommunications*, April 1999, pp. 41–44.

Lucent Technologies Bell Labs Innovations: Cajun™P550™ Gigabit Switch and Routing Switch—Product Information.

Nellist, J.G., and E.M. Gilbert, *Understanding Modern Telecommunications and the Information Superhighway*, Norwood, MA: Artech House, Inc., 1999.

Riezenman, M.J., and W. Sweet, "Communications (Technology 1999 Analysis & Forecast)," *IEEE Spectrum*, January 1999, pp. 29–34.

Rigby, P., "DWDM Orders will Climb Sharply, says study," *FibreSystems*, June 1999, Vol. 3, No. 5, p. 9.

Willner, A.E., "Mining the Optical Bandwidth for a Terabit per Second," *IEEE Spectrum*, April 1997, pp. 32–41.

4

Defense Systems

4.1 Defense—we all still need it

The total global defense expenditure currently amounts to a figure around U.S.$700 billion and could reach one trillion dollars before year 2020—unless some totally unexpected new trend sets in that enables either significant decreases or else substantial increases in defense capabilities.

The approximate defense expenditure of NATO is shown on an annual basis in Figure 4.1. In most years NATO's expenditure contributes around 70% of the global total. The 1990–1993 rise corresponded to the Gulf War in 1991 and subsequent restocking. The 1999 increase is explained by the Yugoslavian War, and the year 2000 drop in expenditure is explained by the anticipated recession driven by the Y2K challenge concerning microprocessors. After the year 2000, expenditures resume the overall inexorable steady increase.

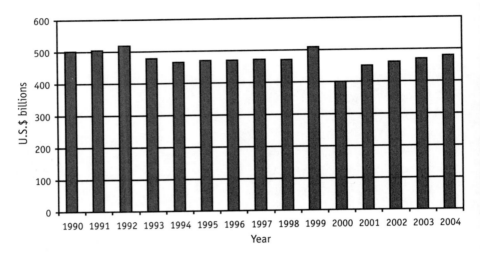

Figure 4.1   Total defense spent by NATO countries.

Most NATO countries are within the G8 group of advanced industrial nations, and their defense technologies are therefore also relatively advanced. As a consequence, much of the defense hardware and software is also advanced. The United States alone accounts for 41% to 42% of total NATO defense expenditure.

During the 1980s, the Falklands War and Iran/Iraq apart, it was almost excusable for anyone to think that after two highly destructive world wars, followed by conflicts such as Korea and Vietnam, a great "world peace" was well on the way. Then, of course, there came the realization that Iraq was amassing arsenals of terrible weapons, including biological, chemical, and nuclear. First the dictator of that country, one Saddam Hussein with his ruling Baáth Party, killed many of his own countrymen (the Kurds) using some of the biological and chemical weapons, and then in 1990 he invaded Kuwait. It was almost as if he were deliberately taunting the Western powers and the United Nations (UN).

There followed, early in 1991, the now notorious Gulf War, and this amounted to the first truly high-tech military operation in history. Cruise and SAM missiles were extensively deployed with devastating effects and apparently incredible accuracy. Late in 1998, following years of UN Weapons Inspectors "weapons forays" in Iraq, a relatively short duration mission known as "Desert Fox" was launched. Actually, due to ever-improving technology, the munitions expended during this campaign were even more precisely targeted than the earlier Gulf War. Techniques such as laser guidance, on-board very high-speed computers, and GPS satellite navigation contributed greatly to the success of these campaigns.

The above describes in outline the highly visible aspects of modern war—the "electronically-directed" deployment of munitions and the radio (including satellite) communications.

There is, however, another extremely important element that characterizes defense and military operations into the third millennium. Arguably this is at least as significant as the military theater itself and here we are talking about the vital surveillance, monitoring, and analysis activities. As any manager knows, without accurate and up-to-date information no task at all can be effectively performed. This is true for both civilian and defense activities, but the main difference is that in the defense instance entire cultures can be altered, often irreversibly, and of course people will die and become maimed if the campaign lacks constantly updated and credible information.

During the final decades of the twentieth century various international agreements, particularly within NATO and what is known as "UKUSA," ensured that these surveillance, monitoring, and analysis activities expanded steadily. Until 1990 this was also true of the United Soviet Socialist Republic, but with the demise of this previous "second superpower," there inevitably came the downfall of the earlier Warsaw Pact security agreement.

Surveillance, monitoring, and analysis are all absolutely essential for the effective pursuit of all modern defense operations. Such activities do not wait until the outbreaks of active hostilities but instead are perpetually under way—there is constant and persistent electronic international security activity occurring at all times. When things "hot-up," as for example with the Yugoslavian situation in 1999, these ongoing security systems just become busier than ever. Later in this chapter some details are provided of the types of communications systems involved in such security installations. Historically such networks derive from what has become known as signals intelligence (SIGINT) or communications intelligence (COMINT).

The total electronics (including all communications) element of defense procurement amounts to about $150 billion globally. NATO accounts for two-thirds of this total, with the United States' defense electronics standing at around $55 billion in 1999 because of the war in Yugoslavia, declining somewhat in the following years. These figures exclude the relatively large research and development (R&D) investments. For example, in spring 1999 the United States Air Force (USAF) announced a budget request for almost doubling the R&D investment in military satellite communications—from $54.2 billion in 1999 to $97.1 billion in the year 2000.

This then sets the scene for the defense communications systems requirements and their scale in terms of typical operations and costs.

It is worth observing that, while cruise missiles travel at only moderate speeds, contemporary fighter jets such as F-16s and Stealth Bombers move at supersonic speeds and often with fractional-seconds time-to-target scenarios. Some "smart bombs" shift at lightning speed, frequently several thousands of miles per hour. Also vast and ever-increasing amounts of data must be transferred between nodes, both mobile and static, in ever-shortening time spans. All these aspects point toward an expanding requirement for very high-speed defense communications systems.

In addition to this high-speed systems requirement, defense communications networks must always be robust and this is reflected in terms of the network topologies. Nodes must be accessible via several different alternative paths or routes. Relatively simple axial or pyramidal topologies, while suitable historically, are unacceptable today and instead matrix or grid forms must be implemented. In Figure 4.2 the upper network, named (a), represents an axial or pyramidal topology where a failure in early links results in total or at least major breakdowns in the entire structure. In contrast, with the matrix network shown in Figure 4.2 (b), signals may become routed over several alternative paths and therefore a failure in one or more paths is overcome (using software control) by the fact the signal has at least one alternative route. In both cases COC means "chain of command."

In practice the various nodes may be within terrestrial, tropospheric, or satellite configurations and many modern defense networks are implemented with appropriate combinations of such approaches.

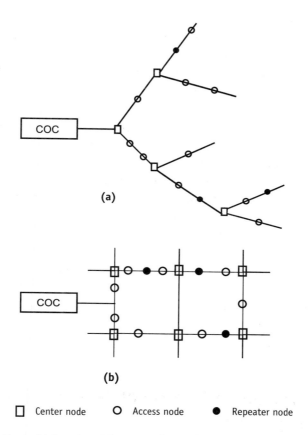

Figure 4.2 Pyramidal and matrix network topologies.

Actual technologies may therefore be copper or fiber cables or they can be implemented using microwave/millimeter-wave carriers.

Obviously with many terrestrial systems, whether static, semistatic, or mobile (land, sea, or air), wireless links using microwave or millimeter-wave carriers are often desired and this is also true for satellite systems. Highly secure cabled links almost always employ fiber-optic cables and encrypted digital signals. This applies to defense as well as many commercial satellite ground station links because the alternative of coaxial copper cabling is inherently insecure due to radiation and vulnerability to tapping.

Some military communications systems incorporate microwave radio, tropospheric scatter, and satellite integrated within one overall network. For example, a typical TSSR (tropospheric and satellite services radio) operates using a 15-GHz carrier on line-of-sight integrating Tr-Tac equipment and GMF (ground mobile forces) satellite terminals.

Tropospheric scatter systems, supplied by corporations such as Aydin, use very high microwave power (multikilowatt) transmitted into earth's troposphere at a narrow elevation angle. Forward scattering of the signal occurs enabling large ranges (hundreds of kilometers) to be covered over challenging terrain. Such systems are generally transportable using medium-sized military trucks.

Digital traffic can be carried, typically up to 6.144 Mbps, using a pseudo non-return to zero (NRZ) signal that is also channeled over fiber-optic cables. For the line-of-sight 15-GHz radios, antenna diameters of either 305 (31 dBi gain) or 610 mm are employed and installation time is generally less than one minute. The system operator chooses either horizontal or vertical polarization for optimum performance over these types of links.

4.2 The MILSATCOM, Skynet, and NATO series of satellite systems

An ever-expanding variety of defense satellite systems and associated spacecraft are being deployed on a global basis. The United States and Western Europe remain the principal sources of space hardware, and nations within Europe, often with companies operating in consortium-like groups, are in the forefront of this technology.

For example, the Netherlands' Ministry of Defense planned to complement its telecommunication systems with satellite communications and an official budget was allocated in the late 1990s for this extension. As a first step the Netherlands' government sought cooperation in Europe to join or to come to a bi-, tri-, or multinational satellite system. Under an agreed arrangement the Netherlands would become the supplier of important subsystems as a contribution in kind to satellite systems planned by other countries. A participation of the Netherlands in the definition phase of a Military Satellite Communication (MILSATCOM) program was considered essential both for the Ministry of Defense and the Netherlands' industry.

The Netherlands Agency for Aerospace Programs (NIVR) coordinated and promoted, in consultation with others, the capabilities of the Netherlands' space and ground segment industry. The Netherlands' industrial MILSATCOM space-segment team is led by Fokker Space, and the ground-segment team was led by Fokker Aviation and Signaal Communications. It is important, however, to note that the term MILSATCOM is by no means confined to any one specific program such as this Netherlands project. Other systems, including Skynet, are also brought under the MILSATCOM umbrella term.

A schematic outline of a MILSATCOM spacecraft is shown in Figure 4.3, dominated as usual by the solar energy collecting panels. All of the on-board electronics including the on-board signal processing are contained within the "PSU and Transponder Unit" (PSU is the abbreviation for power supply unit) and the dish reflector antennas focus the uplink and downlink microwave beam directions as required for differing theater operations.

In defense terms *Skynet* refers to a sequence of military satellite systems implemented by the U.K. Ministry of Defence since the 1970s—starting with Skynet 1. British Aerospace (B Ae Space Systems) was the prime contractor for Skynet-4, and the Franco-British joint venture known as Matra Marconi Space was responsible for the communications module. A schematic illustration of a group of Skynet ground stations, based upon a photograph of the RAF (U.K.) Colerne installation, is provided as Figure 4.4. These large, high-gain, dish reflector antennas provide secure communications links for U.K. armed forces and the NATO alliance. The final three Skynet-4 spacecraft were being launched over the 1999–2000 time frame and a new system was being planned.

For the latest generation Skynet system (Skynet-5) a radically different contractual approach was proposed in 1999. It has been proposed that Skynet-5 should be developed under a private-public partnership arrangement in contrast with the traditional procurement program and a two-year study was set up in the spring of 1999 with the aim of awarding study contracts to two competing teams. The first team, which might be called the "British-British/U.S.-U.S. Team," comprised British

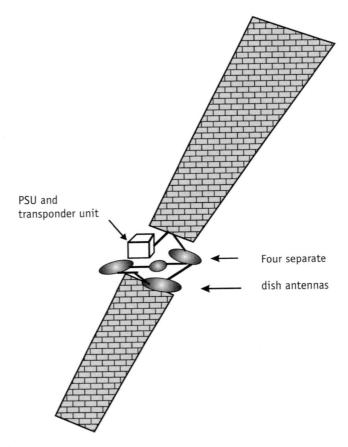

PSU and
transponder unit

Four separate

dish antennas

Figure 4.3 Outline of the MILSATCOM spacecraft.

Figure 4.4 Skynet ground station schematic.

Aerospace (BAe), British Telecom, TRW, and Lockheed Martin—that is two British firms and two U.S. corporations with a balanced expertise and experience in appropriate fields. The second team, named Paradigm Secure Communications, comprised just two companies: Matra Marconi Space and, again, TRW.

This was the first time that any government had proposed privatizing a classified satellite program, and doubtless the maintenance of secrecy is the principal challenge facing both government and contractors during all phases including the study and the final production contract. Several factors influence this approach: cost savings to government and therefore the taxpayer; opportunities for industry during the study, development, and production phases; and the possibility of trading any excess satellite link capacity to commercial customers and/or NATO. Industrial investment to the tune of approximately U.S.$5 billion is anticipated.

Like the advanced, extremely high-frequency (EHF) MILSTAR replacement system proposed by the U.S. Department of Defense for 2006, Skynet-5 would also include an EHF communications element. EHF means the use of frequencies above 30 GHz and substantially improved security is a main advantage perceived by the defense organizations as a consequence of such an extremely high-frequency carrier. The Skynet-5 development and production contract is scheduled to be let in 2001 and the in-service date for the system is expected to be 2005. In addition to the EHF communications package, mainly for the use of the U.K.'s nuclear forces, there will also be an X-band (8–12 GHz) communications element to service tactical units.

In the tactical scenario, rapid setup is a vital requirement and the narrow beamwidth focus associated with EHF antennas is unacceptable so that a lower carrier frequency is essential in this case. This forced the decision to adopt X-band for the tactical units use.

Since about 1980 many highly transportable types of military satellite terminals have been announced, mainly by U.K. and U.S. manufacturers. Two examples from the United Kingdom are cited: the military off-the-shelf satellite terminal (MOST) and the MARMOSET 1.7-meter tactical antennas manufactured by Matra Marconi Space U.K. (hence the acronym).

The MOST system comes in two pieces: a 2.4m automatically tracking and motorized dish antenna and a communications module.

Transmission is via a 90W all solid state X-band power amplifier and a receiver low-noise amplifier is also housed within the same unit. Data can be transmitted and received at rates up to 2 Mbps and assembly-disassembly time is from 1 to 2 hours. MOST is specified as being suited to either tactical deployment or for semipermanent network hub situations.

The MARMOSET unit is aimed specifically at tactical applications and, apart from the antenna itself, much of the required RF electronics can be attached to the body of the support tripod provided. This antenna is hand-wheel-steerable for aiming at the desired spacecraft.

A North American company known as Satellite Transmission Systems (Hauppage, New York) manufactures what is termed a "Fly Away Satellite Terminal" (the FAST 1A). This is designed to provide rapidly deployable and mobile voice, data, and compressed video communications. The uplink frequency range is from 14 GHz to 14.5 GHz and downlinks can be established within any of the following frequency ranges: 10.95–11.7 GHz, 11.7–12.2 GHz, or 12.25–12.75 GHz. Satellite Transmission Systems provide options regarding power amplifiers ranging from 5W to 300W so that the users can choose the lowest workable power depending upon the exact circumstances.

Either local or remote controlled options are also available. FDMA/SCPC (single channel per carrier) is used and data rates can reach 23 Mbps.

Basic voice and data communications between semimobile ground forces and satellites can be accomplished using relatively small, simple, and lightweight antennas feeding equally small and lightweight electronics units. There is an increasing emphasis on the adoption of flat-plate antennas rather than the traditional parabolic dish reflectors. Typical flat-plate antennas have square dimensions around 30-mm sides and approximately 12-mm thick. A "matrix" of dielectric coating material protects the underlying antenna array in two senses: in terms of environmental protection and also in terms of security (the enemy has a hard time determining much about the actual antenna—its likely frequencies of operation, and so on). Often there are four subsidiary panels that are demountable so that one person can assemble or disassemble the unit in less than 30 seconds.

Behind each square panel there are many metal antenna array elements that are all driven electronically by the requisite satellite signals. Rockwell Collins' DATA*SAT unit is typical of this type of system.

Aboard a naval frigate, for example, all deck-based equipment must be extremely rugged to withstand the rigors of the weather and sea states. Therefore naval SATCOM antennas are almost always mounted on rugged support structures. Typical examples are manufactured by Raytheon E-Systems and such units are compatible with DSCS (defense security—described below), HISPASAT, NATO SHF, Skynet, and also Syracuse systems. These typically C- or Ku-band terminals provide continuous operation with automatic tracking under all conceivable sea state conditions.

The NATO organization requires its own satellite systems for the integration of the SATCOM element into the NATO Integrated Communications System (NICS) and the history of the program goes back to the 1960s. The spacecraft are of similar configurations to that of the United Kingdom's Skynet series of spacecraft—at least up to the existing Skynet-4 generation.

A NATO IV spacecraft is illustrated in Figure 4.5. The rigid solar wings, PSU, and transponder units are clearly visible. The transponder unit is also frequently termed the *communications module* and this contains the communications payload mounted beneath the service module. This payload comprises:

- A three-transponder SHF package—each transponder having 40W transmit power and between 60-MHz and 135-MHz bandwidths;

- A UHF package that has two 25W transponders each of which serves a 25-kHz channel;

- Anti-jamming (EW) electronics in addition to the usual signal processing facilities.

The SHF (microwave) transmit and receive antennas provide for either substantial earth surface cover or relatively narrow "spot beams" as required. These are the dish reflectors on the side of the transponder unit. In contrast, the UHF antenna is the helical unit projecting from the

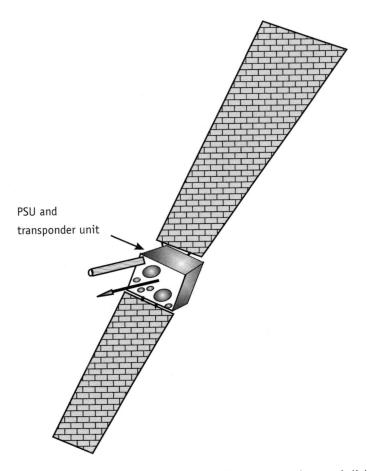

PSU and
transponder unit

Figure 4.5 A NATO IV spacecraft; a large spiral antenna and several dish reflections are shown.

module. This UHF antenna provides extensive earth surface coverage but at the expense of relatively restricted bandwidth.

It will be interesting to see in the future as to whether any new configuration (NATO V?) will again follow the new Skynet-5 design and contractual philosophy as described above.

Defense Satellite Communications System (DSCS) configurations are described in a later section.

4.3 Military microwave radio relays

Microwave and millimeter-wave radio relay systems designed for military applications have much in common with their civil and commercial counterparts. There is always the basic option of analog or digital modulation and multiplexing, the behavior of the signals through space is identical, and either solid state or vacuum-tube-based power amplifiers are available depending upon the power levels and attendant frequency ranges.

The principal distinctions arise from the specific defense and military need for high reliability, security, and low probability of intercept. High levels of encryption ensure high security and the adoption of the millimeter-wave bands (EHF) enables the use of small-size antennas with high directivity. Such antennas yield intrinsically low probabilities of intercept due to their narrow beams and low sidelobe levels.

As described above in relation to EHF satellite links, although millimeter-wave radio relay systems are inherently terrestrial and therefore of relatively short range, the shift to these much smaller wavelengths results in more critical setting up in terms of pointing accuracy. Fully automatic antenna pointing is therefore highly desirable.

An example of a digital millimetric wave radio relay is the MH938 product, manufactured by Marconi Communications (Genoa, Italy). This equipment operates within the 37–39.5-GHz band, well within the millimeter-wave ranges. Traffic up to 34 Mbps is handled and other characteristics include:

- Main channel:

  - 0.5 to 0.6W output power at the antenna connector;
  - CP-FSK modulation;
  - Information rates to 34 Mbps.

- Auxiliary channel:

  - 0.5W approximate output power;
  - QPSK (quadrature phase-shift keyed) modulation;
  - Either 1×64 Kbps or 3×64 Kbps data rates.

It is noteworthy that the antenna comprises a lens unit that is completely sealed, environmentally protected, and shielded to ensure negligible sidelobe or other stray radiation.

4.4 Other defense systems

As described earlier in this chapter, surveillance, monitoring, and analysis are all absolutely essential for the effective pursuit of all modern defense operations. Also it is clear that these types of defensive activities cannot wait until active hostilities begin but instead must be operated on what is essentially a continuous basis. At particularly critical times, such as the Yugoslavian situation in 1999, these ongoing security operations become exceptionally busy. The traditional concept of this type of operation amounts to signals intelligence (SIGINT) or communications intelligence (COMINT), but modern covert operations are extremely sophisticated.

A specialized series of satellite systems known as the Defense Satellite Communications System (DSCS) continue to be deployed to provide communications services for the U.S. Department of Defense (DoD), the National Security Agency (NSA), the White House Communications Agency (WHCA), NATO, and the United Kingdom. An example of a DSCS series III spacecraft is shown in Figure 4.6, enhanced by the addition of an EHF wideband payload. This type of spacecraft is equipped with the following antenna arrays: 19-beam SHF transmit, 61-beam SHF

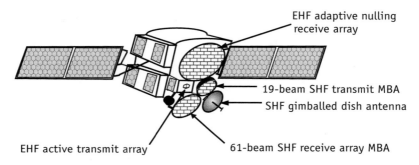

Figure 4.6   A DSCS III spacecraft enhanced with an EHF payload.

receive, an SHF gimballed dish reflector, EHF active transmit, and an EHF adaptive nulling receive array (MBA stands for main beam array). These facilities are generally considerably more sophisticated than those provided for most other defense-orientated spacecraft.

The Airborne Early Warning AirCraft Systems (AWACs) represents another important NATO entity. In 1999 NATO equipped its Main (AWACs) Operating Base (MOB) at Gelienkirchen in Germany with a simulcast radiocommunications network founded upon what is known as "Tetrapol" digital technology.

This network was designed to provide secure communications as well as close coordination for all staff at the base, including emergency situations. Included are the military command, the control tower, all emergency services, and also maintenance teams. Daimler Chrysler Aerospace (DASA) of Germany was a main contractor and the network cost was estimated as approximately $1.6 million. In addition to the control tower installation, fire alarm center, and the dispatch nodes, there are 200 handheld units in addition to some tens of vehicle-mounted and fixed terminals.

4.5 Global security systems

The National Security Agency (NSA) was gradually developed by the United States during the years of the cold war, the 42-year standoff between what were perceived by most to be the world's two superpowers of those times—the United States and the United Soviet Socialist Republic. With electronic intelligence gathering (ELINT) being widely recognized as critical in modern warfare—both cold and hot—it was completely understandable to find the protagonists evolving sophisticated global eavesdropping facilities.

By the late 1980s the NSA had grown to become America's most secret, largest, and almost certainly most costly security organization. The NSA has access to land-based, ship-borne, and air-based facilities and uses satellites together with other communications systems to channel information back to its intelligence bases. Associated mobile security systems are located on board ships, submarines, aircraft, and spacecraft globally.

The NSA has built up intelligence links with the following four close allies: Australia, Canada, New Zealand, and the United Kingdom. The resulting secret agreement is called UKUSA and all five flags of these nations are frequently displayed at each UKUSA location—and definitely on days of special VIP visits. UKUSA refers to the high-level official agreement between these close allies to exchange intelligence information on what amounts to a continuous basis.

The NSA at Fort G. Meade in the United States is Anne Arundel County's largest employer and it is also one of the largest employers in the entire state of Maryland. This organization implements information encryption techniques that are second to none internationally, and the term *cryptologic*, referring to cryptographic systems, is frequently used. More standard terms include SIGINT, meaning (electronic) signals intelligence, and also INFOSEC, which is shorthand for information security. Most buildings located in the NSA facilities are windowless for additional security. With walls well shielded against "stray" electromagnetic radiation, using internal metal coatings, and the windows themselves unable to let radio or optical signals either in or out, a high level of security is essentially guaranteed.

All forms of electronics, including very high-speed digital communications systems, are of vital significance to the NSA and also essentially all other security organizations.

In order to receive sensitive low-power signals from many sources quite large dishes are essential, covered with white radomes. The radomes protect from the worst effects of bad weather which cause signal loss, help keep the important parts of the dish clean, and also stop any onlooker from seeing the direction in which the antenna is pointing—important where security is paramount.

The radomes are usually of a neutral white outer surface color. White reflects the sun's rays very efficiently and limits the extent to which the system heats up. On the whole, the hotter the satellite receiver, the worse the reception quality because background noise increases. In all probability the security people would prefer their radomes to be camouflaged so as to merge with the natural surroundings, but in the end a white surface wins out.

By the year 2000 the Menwith Hill NSA base enjoyed the benefits of some 29 radomes in total, with several more probably still to be added.

An elevational view of the main sections of the base is provided by Figure 4.7, which is founded upon an actual photograph published in the press during the late 1990s. Several differently-sized radomes are clearly visible as well as windowless buildings and other structures. The base is surrounded by the open countryside of North Yorkshire County, England. According to well publicized media reports, entry into the most recent facility at Menwith is controlled by a retinal imaging technique in which only those staff for whom matching (stored) retinal images precisely correlate gain access.

An aerial view, again based upon an actual late 1990s photograph, is given in Figure 4.8. There are some additional radomes that are not visible in this particular view. Areas comprising collections of many small buildings are indicated to the left background—these are mainly administrative together with some residential accommodation. Further rectangular blocks house the operations elements of the base and these are much closer to the radomes. A relatively insecure dish reflector antenna, without the benefit of a radome, is visible in the foreground.

The radomes, in any operational base, are by no means randomly located. Some groups of radomes are especially interesting because they appear to comprise large phased array configurations. In Figure 4.8 two such groups are identified: group A, B, and C; and also group i, ii, iii, and iv. The radomes within each of these groups fall into distinct lines—that is they form linear rows and it is well known that these types of configurations usually amount to phased arrays. The best known nondefense example in the world is probably the Very Large Array (VLA) in Socorro, New Mexico (U.S.), that is operated by the National Radio Astronomy Observatory and which comprises a total of 27 individual antennas mounted on railroad tracks.

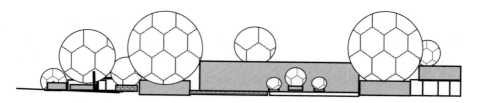

Figure 4.7 A westerly elevational view of the Menwith Hill Base (based upon an actual photograph).

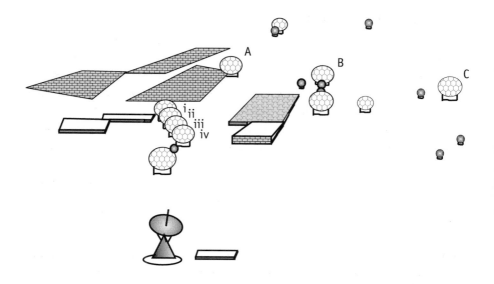

Figure 4.8   An aerial view of the Menwith Hill Base (antennas and buildings—from an actual photograph).

With phased arrays the individual antennas, within the radomes, are electronically interconnected and the signals are fed to and from each antenna with differing respective phases—hence the term *phased array*. Using this principle the resultant composite microwave beam, associated with all the elements of the array, is not only combined—it can also be (extremely rapidly) electronically steered so as to point in different directions. This would have considerable advantages for satellite-based surveillance, including the situation whereby it is impossible to know pointing directions without any knowledge of the electronic steering programming.

In phased arrays the separation of the antenna elements determines the resolution of the final scanned image—the larger the separation the greater this resolution—so array A, B, C should have a better resolution than i, ii, iii, iv. It is also interesting to note that the two distinct arrays are disposed approximately 45 degrees from each other. This probably means that they track quite different satellite constellations. Figure 1.16 of Chapter 1 indicates the basic concept.

Menwith is a key element in the now expanding Star Wars program. This program embodies space-based infrared systems (SBIRs) and high-power laser guns triggered by remotely (from earth) controlling selected groups of satellites. General Chuck Horner, the retired Desert Storm (Gulf War) commander, is on record for stating that the major aim of the Star Wars program is "neutralizers in space"—a means for neutralizing enemy missile attacks.

During the late 1990s a further $5 billion was earmarked by the U.S. Congress for this Star Wars "Space Command."

4.6 Typical technologies

Most of the basic technologies associated with defense installations today are essentially commercially available. It is the final systems configuration—the ultimate systems specification—that determines the defense nature of the installation.

The space segment represents the exception but this should not be surprising in defense application terms because components and modules fitted for space must always be space-qualified. In many instances such components must be "rad-hard," that is, they must remain operational in environments where severe ionizing radiation flux is present, up to a specified limit.

In most current and projected applications solid-state components and modules are implemented in the bulk of the systems. However, where relatively high power is required, particularly at higher microwave and millimeter-wave frequencies, tubes such as traveling wave tubes (TWTs) remain essential components. In some systems other types of tubes, including klystrons and magnetrons, are still encountered. Certain narrowband satellite and also tropospheric scatter transmitters necessarily employ multikilowatt klystrons.

The power capabilities of solid-state components and modules are, however, subject to continuous improvement resulting from new device technologies and advances in power-combining techniques. Monolithic microwave integrated circuits (MMICs) represent a very significant aspect of this scenario. In particular, the Defense Advanced Research Projects Agency (DARPA—under the Department of Defense) frequently reports on the status of such elements. The situation that applied to MMICs in the late 1990s is indicated in Figure 4.9, and to achieve output power levels exceeding 100W solid-state MMICs are used as drivers for tubes such as TWTs. In this case the signal is first amplified by the MMIC and the output from this MMIC is then applied to the tube input. Typically, a MMIC output power level of 0.5W at 26 GHz is followed by a tube output power of the order of 500W at the same frequency.

The upward-pointing arrows on the power-frequency trend line refer to the fact that power capabilities tend to increase, although certainly not indefinitely. As operating frequencies move inexorably into the millimeter-wave portion of the spectrum, so the availability of power above 1W from individual MMICs decreases substantially. Power combining must then be used in order to obtain the desired increases and spatial, rather than "on-slice," combiners are gaining acceptance because the overall efficiency is maintained almost regardless of the number of individual MMICs.

In this technique up to several hundred MMICs may have their individual outputs "funneled" into one final output port.

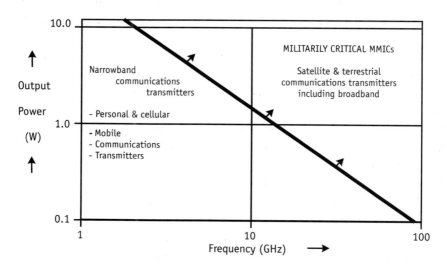

Figure 4.9 MMIC power output trend.

Since the mid-1990s the important concept of commercial-off-the-shelf (COTS) component procurement has been an official and vital aspect of the industry. COTS embodies a range of different levels and it is important that components and subsystems are designed with the appropriate level of specification compatible with the final system destination. Clearly the specification applying is much higher for a fighter aircraft, for example, than for a defense headquarters building.

There exists a Qualified Manufacturers List (QML) with an attendant performance base specification (MIL-PRF-38525) and also a Qualified Parts List (QPL), bearing standard specifications MIL-N-38510 and MIL-STD-883.

The commercial capabilities of firms manufacturing products in accordance with QML are epitomized by, for example, ceramic-packaged "small outline ICs" (SOICs) that are manufactured by firms such as Analog Devices and National Semiconductor. Such corporations supply multichip modules (MCMs) implemented using SOIC technology, and increasingly these MCMs are of the mixed-signal type. This means that many chip functions are contained within the same MCM—from millimeter-wave input signals bearing Gbps data rates through to amplifiers and digital signal processing chips.

4.7 Summary

Taken overall the $700 billion (annual) global defense industry is massive by any standards. Of this $700 billion at least $150 billion is due to electronics and communications systems.

Electronic warfare (EW) is of vital significance and almost always demands broadband operation. Tropospheric and millimeter-wave terrestrial radio are important, as are satellite systems such as the MILSATCOM, Skynet, and NATO series. Apart from war theater operations, global security organizations such as the NSA are substantial and almost continuous consumers of defense-type systems.

Underlying technologies include flat-plate antennas, covering radomes, phased arrays, MMICs, and vacuum tubes for high power output transmitters.

Human nature will continue to ensure that, however unfortunate it may seem, defense operations and expenditure will be with us practically indefinitely.

Select bibliography

Defence Analytical Services Agency (DASA-UK), *The Resource Context: A Comparison of Defence Expenditure: NATO Countries.* (Annual data).

Janes Military Communications, 1998, Janes Books (numerous references).

Sagan, C., *Cosmos*, Carl Sagan Productions, Inc., 1980, pp. 260–261.

"Shared Radio Spectrum Enhances Communications at Key Airbase," *Land Mobile*, March 1999, editorial item, p. 10.

"Skynet-5 Project will be Testbed for Funding Study," *Janes Defense Weekly* (JDW), March 24, 1999, editorial item, p. 30.

"Space Research and also Signals Intelligence," *Janes Defense Weekly* (JDW), January 27, 1999, editorial item, p. 8.

5

Digital Television

5.1 Convergence revisited

Today almost everyone agrees that digitization is the way to go as far as signal transmission and processing are concerned.

While all "real-life" signals are initially strictly analog, the digital encoding of these signals at the earliest possible point well ahead of transmission provides immense benefits. This concept makes sense for several compelling reasons, for example:

- Minimization of noise effects;

- Enabling of substantial bit-compression of video and audio signals;

- Enabling error correction to be deployed (which is not feasible with analog);

- Digitization means that identical processing techniques can be used for all signals, whether audio, control, data, or video;

- Encryption is easily arranged;

- Channels can be easily allocated;

- Lower (than analog) transmission power is usually required;

- Robust modulation schemes, combating ghosts and phase problems, may be implemented;

- The implementation of truly high definition displays (HDTV) makes much more sense than with analog;

- Compatibility with most of the burgeoning digital technology that is generally available (e.g., VLSI technology);

- Compatibility with many of the increasingly digital transmission systems that are being implemented in almost all G8 countries for at least cable and satellite.

By the late 1990s these advantages were realized with the introduction of digital television (DTV) in several countries. DTV was initially pioneered with satellite (DTH) but has since become possible using terrestrial broadcast TV mainly as a result of the advances in digital compression technology.

There are, however, some drawbacks to implementing digital TV and the main ones are:

- The quality of service degrades rapidly as the receiver departs from the edge of the service area.

- Subscribers must at least purchase a "set-top box" (described shortly)—or else an entirely new set.

- Terrestrial broadcasting stations need to develop a new transmission infrastructure.

With satellite DTH no changes to the infrastructure are required.

We have already discussed the practically complete convergence between computer technology and communications, in both hardware and software terms, and this can be seen most vividly by observing the Internet explosion. However, not everyone wants to explore the

Internet all the time and this will likely continue to be the case well into the third millennium. Meanwhile broadcast or cabled (CATV) radio and television programs remain of great significance to many for movies, sports features, and news. Quality of reception, reliability, and the facility of effective program dial-up will always attract subscribers. The highest conceivable reception quality, compatible with wide screen viewing and highly flexible program key-request, is available only with DTV.

So, given the pervasiveness of computers (PCs), why did it take until the final eighteen months of the twentieth century for DTV to finally catch on?

With radio and television the principal problem is history—history that many of us still live with. Historically radio and TV broadcasts have remained defiantly analog. Modulation techniques such as amplitude or frequency modulation (AM/FM) and vestigial sideband (VSB) have steadfastly remained in common use.

Well into the third millennium it's a reasonably safe bet that most radio receivers, portable or otherwise, and also many television receivers, will still be analog even in the so-called advanced economies of the world. In third-world nations and most emerging economies, domestic sets will almost all remain analog.

As far as DTV is concerned a variety of standards have emerged which tend to reflect national (or locally transnational) industry and its protection. There are distinct differences between Europe (including the United Kingdom) and the United States, although Japan has adopted a similar system to that of the latter. Australia, however, has opted for a system that differs substantially from others elsewhere in the world. Where are global standards!

The largest potential market is China where the decision of the massive domestic TV company called China Central TV is critical.

5.2 Technologies now—and then

Clever and often highly advanced technologies were applied to the radio and television systems of the twentieth century—almost all of them analog. Subsystems such as wideband voltage-controlled oscillators, linearized frequency modulators, and phase-sensitive detectors have all been developed and placed into volume manufacture at various times for analog systems.

Interestingly, with the advent of DTV, much of this conventional technology will remain significant for several years to come. This is because in principle a broadcast or cable subscriber can receive DTV in one of two possible ways:

- By having a completely new integrated receiver system;

- By installing a set-top box either literally on top of or close to the analog TV set.

Obviously a completely new receiver system will embody all the requirements for efficient reception at the same time as providing a high-definition display. The DTV signal enters the subscriber's premises via either cable or the antenna and proceeds directly into the receiver. However, in many cases the digital signal will enter the subscriber's premises either from cable or the broadcast receiving antenna and will still proceed to the immediate environs of the conventional receiver. At this point the DTV signal is ported into the set-top box or "outboard decoder," a concept very familiar to CATV subscribers for many (analog) years. This decodes the signal into a form suited to the processing capabilities of the conventional analog set.

The options are illustrated in Figure 5.1. In these two schematic illustrations the antennas and the associated coaxial cable downloads (shown as curved lines) are identical in each case. With case (a) the TV is a dedicated DTV receiver, and in case (b) the set is a conventional one. The sole purpose of the set-top box is to convert the incoming high-quality digital TV signal either from terrestrial or from a satellite (DTH) transmission into an equivalent analog signal as required for the conventional analog TV set.

With the set-top-box-into-analog-TV approach several of the major advantages with DTV are lost—essentially thrown away—because of course all the characteristics of the conventional analog TV remain in place. These include lack of the potential availability of truly high-definition displays and no internal compatibility with most of the

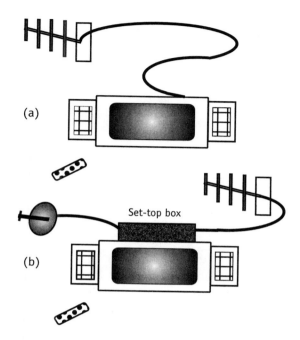

Figure 5.1 DTV reception: (a) to DTV set and (b) to set-top box on conventional set.

burgeoning digital technology. If the design is rationalized correctly the same set-top box can include both DTH and DTV.

The various formats associated with DTV are important and they logically lead to the required precompression transmission bit rates. The five major ATSC (Advanced Television Systems Committee) formats are identified in Table 5.1.

In this table a compression ratio of 91:83 has been assumed in accordance with the MPEG2 standard.

Each bit rate refers to a particular frame rate (frames per second) for the TV signal. For example in the SDTV 480p (4:3) case, 295 Mbps is associated with a 60 frames per second rate, 148 Mbps with a 30 frames per second rate, and 118 Mbps with a 24 frames per second rate. This logic continues through all the cases cited. For HDTV 1080i no 60 frames per second rate applies, and for the 1080p standards are not yet fully defined.

Table 5.1

Major ATSC Television Formats

Format	Name	Screen Ratio	Scan	Resolution (Pixels)	Precompress Bit Rate (Mbps)	Postcompress Bit Rate (Mbps)
SDTV	480p	4:3	Progressive	640 × 480	295/148/118	3.21/1.61/1.28
SDTV	480p	16:9	Progressive	704 × 480	324/162/130	3.53/1.73/1.42
HDTV	720p	16:9	Progressive	1.280 × 720	885/442/334	9.64/4.81/3.64
HDTV	1080i	16:9	Interlaced	1,920 × 1.080	NF/995/796	NF/10.8/8.8
HDTV	1080p	16:9	Progressive	1,920 × 1,080	Not yet known	Not yet known

It can be seen from Table 5.1 that the formats range from standard definition TV (SDTV) at only 640 × 480 screen resolution (i.e., corresponding to the most basic VGA computer monitor format) up to the high definition TV (HDTV) at a resolution of 1,920 × 1,080 pixels. This highest level format well exceeds the enhanced computer standard of SVGA at 1,280 × 1,024 pixels.

In terms of total pixels, which represent a factor often used when comparing formats, both 1080 HDTV formats lead to over two million pixels on screen (2,073,600 to be precise). By comparison the most basic standard, SDTV 480p, only yields just over 300,000 pixels (307,200). True high-quality HDTV demands the over 2-million pixel count standard.

In the compression process redundant picture data is neglected and static or very slowly changing blocks of pixels are simply coded to repeat through frames until a change is detected. Data is, however, provided concerning shadows caused by the movement of eyes, lips, hands, and other limbs. In this way a dramatic reduction in transmission bit rate is achieved.

During the late 1990s the standardization of a number of international video coding schemes was announced. These included ITU-T H.261 and H.263, and ISO/IEC MPEG1 and MPEG2, and all of these addressed a large range of applications with different requirements, for example, in terms of bit rate, quality, or delay.

A relatively new standard, known as MPEG4, aims to provide a universal, efficient coding of different forms of audiovisual data generally termed *audiovisual objects*. This basically means that MPEG4 intends to represent the world in terms of a composition of audiovisual objects, following a script that describes their spatial and temporal relationship. This type of representation should provide the opportunity for the user to interact with the various audiovisual objects in the scene, in a way similar to the actions taken in everyday life.

For example Matsushita Electric, branding under the Panasonic and National names outside of Japan, has taken up the MPEG4 video compression standard. This corporation filed 32 patents and claimed that this approach is twice as efficient as MPEG2, which should mean for instance that the 955 Mbps compresses to 5.2 Mbps. A fundamental feature of MPEG4 is the transmission of distinct specific elements of any video scene. One layer would be the background, another perhaps still or slowly moving objects, yet another relatively fast movers, and so on. In this way identical source material may be decoded employing either broadband DTV or even mobile videophones.

The fundamental technique is termed *adaptive bit technology* that dynamically allocates available resources in layers defined in terms of motion, shape, and texture. A still or only slowly changing layer is allocated more bits than a rapidly moving portion of the overall scene where errors are less detectable to the eye.

There are three main classifications of HDTV receivers:

1. Rear projection (the main type);

2. Direct view;

3. Plasma screen.

A photograph of a direct view iDTV (Philips) is provided later in this chapter.

Most HDTV receivers are of the rear projection class and screen diagonal dimensions can be as high as 73 inches which is equivalent to just over six feet or 190 mm (e.g., Mitsubishi). Direct view screens are by definition of smaller dimensions, with 34 inches or 85 mm being the norm. Some flat panel plasma screens are also coming on to the market and these tend to use the otherwise very unusual (for TV) XVGA format. Sony and Thomson have been leading the way with this technology.

Integrated widescreen HDTV sets are supplied by a variety of manufacturers including Daewoo, Hitachi, Konka, Philips, Samsung, Sharp, Sony, Thomson, Toshiba, and Zenith. Stand-alone widescreen HDTV-ready displays are available from Mitsubishi, Panasonic, and Pioneer. Any of the foregoing types of products, whether integrated or stand-alone, were priced upwards of $5,000 (Daewoo) in 1999.

At the end of the twentieth century Japanese manufacturers had a stranglehold on the set-top box or "outboard decoder" market with Thomson standing alone as the only non-Japanese supplier with the lowest-priced set-top box ($150). This includes the receiving satellite dish but it relies upon a subsidy from the service operator. Otherwise such products were generally priced around $1,500 to $3,000. Most of the sets have satellite TV (i.e., DBS) tuners already built-in.

In common with most technology-sector products, the unit prices are almost certain to fall with time and likely trends are indicated in Figure 5.2. The decrease is relatively steep over the 1997–2007 time

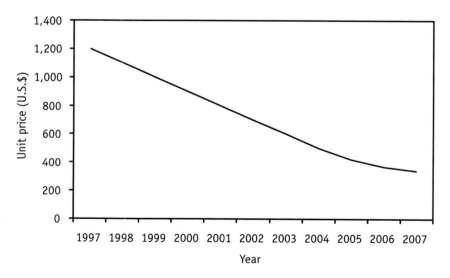

Figure 5.2   DTV set-top box price trend.

frame, as more and more set-top boxes enter the market in volume to meet DTV demand at an affordable price compared with dedicated HDTV sets. Thereafter the unit price essentially falls only slowly—in the low hundreds of dollars range.

A typical set-top box is illustrated in Figure 5.3 and a Philips iDTV set is shown in Figure 5.4.

The choice of modulation scheme is always important in communications and DTV is no exception. Several schemes have been implemented, in particular:

- BPSK (binary phase-shift keying);

- QPSK (quadrature phase-shift keying);

- TC8PSK (trellis-coded quarternary phase-shift keying).

Trellis encoding involves adding the coding information to the conventional modulation scheme (e.g., QPSK or eight-level PSK) to result in a signal characterized by successive values combining the coding itself with the conventional modulation scheme. A "reverse tracking" approach is employed whereby the previously received signal is compared with the more recent signal. In this way a scheme such as TC8PSK has a greater tolerance to noise than either BPSK or QPSK. Figure 5.5 shows a partial vector diagram of the trellis-coded interrelationships between four binary states. In practice there are four further states but only the central four have been included in Figure 5.5, for relative simplicity.

The appearance of the interconnections indicates how the name *trellis* became coined for this modulation scheme.

Figure 5.3   A typical DTV set-top box. (Courtesy Philips Electronics.)

Figure 5.4   An iDTV receiver. (Courtesy Philips Electronics.)

Adopting TC8PSK enables the information transmission capacity to be one-third higher than available with the other more traditional schemes within the same bandwidth. Unfortunately the susceptibility to rainfall is increased but techniques are available to counteract this drawback. It is also necessary for a trellis-coded communications channel to be quasilinear in a similar manner to that associated with a vestigial sideband (VSB) local broadcast channel.

Using TC8PSK information bit rates as high as 50 Mbps can be handled with associated received carrier-to-noise power ratios of 10.5 dB (27-MHz bandwidth), provided the rain attenuation is not much above 4 dB. QPSK, the next best scheme, is limited to 40 Mbps although 7 dB of rainfall attenuation can be tolerated at this bit rate.

Multipath received signals, where there is more than one viable signal path towards the receiver, represent the most serious potential problem

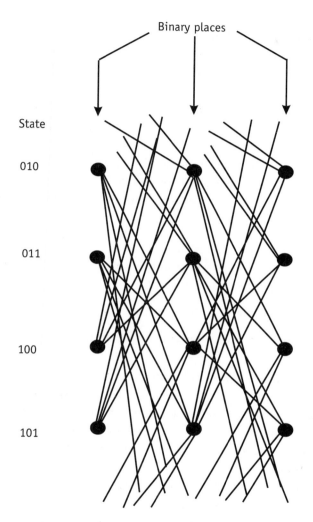

Figure 5.5 Partial "vector" diagram of trellis TC8PSK code.

in DTV. To overcome this, the DTV sets are designed with internal circuitry that analyzes the incoming multiple signals and automatically identifies the most robust of these. This is a standard process that is well known in the industry and is generally termed *adaptive equalization*.

This has, however, already resulted in considerable problems for the industry—especially in the United States. Here 8-bit vestigial sideband (8-VSB) modulation was first adopted for DTV but this is not particularly robust against multipath. The European standard, known as coded orthogonal frequency division multiplexing (COFDM) is by contrast much more robust.

Frequency-division multiplexing (FDM) was commonly used in communications of many forms before the digital age led to a much greater emphasis on time-division multiplexing (TDM). The principle of FDM is basically very simple: the radio spectrum is carved up into a series of adjacent channels or spans of frequencies and different programs are transmitted within specific channels. There are, however, problems with this basic technique, notably:

- It is necessary to introduce "guard bands" between the actual (information-bearing) channels in order to avoid overlap and consequent adjacent channel interference.

- There is essentially no inherent protection against multipath.

- Much precious radio spectrum is occupied.

The last two aspects are of considerable significance, and COFDM has been developed largely to overcome these two problems. With this scheme the signal is spread over thousands of narrow (FDM) channels each of which operate at relatively low information rates. Each of the channels is coded and those portions of the signal that are to be transmitted through specific channels have the appropriate code attached such that no signal elements can become lost. In this way the receiver can efficiently identify and reject undesired reflected (multipath) signals and only enable the strongest desired signal to be processed further through the set.

Frequently termed *DVB-T*, this COFDM approach can be used with the Digital Video Broadcasting (DVB) standard and the bandwidth required is typically around 7 MHz.

The Japanese also use a form of this approach and they have shown that a single frequency network can be supported in this manner. NHK Science and Technical Research Laboratories have been in the forefront, and they have found that the signals are much more resistant to ghosts and multipath fading.

Finally, precious radio spectrum is saved because many services can be supported by a single radio frequency network.

A block diagram of a typical COFDM terrestrial DTV receiver is shown in Figure 5.6.

The TV signal usually enters the antenna and the tuner input at UHF—ultra high frequency, that is, the 300-MHz to 3,000-MHz (or 3-GHz) range—and for signal processing this carrier frequency must be substantially reduced. AGC refers to automatic gain control, which ensures that the signal strength is maintained approximately constant where necessary. The filtered output of the tuner is the first intermediate frequency (IF1) and the output of the next superheterodyne section (the mixer) is the second intermediate frequency (IF2). These

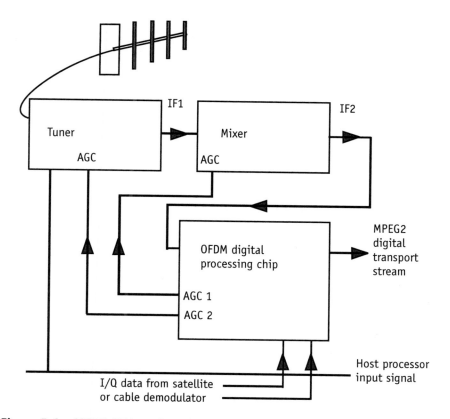

Figure 5.6 COFDM DTV receiver signal processing.

frequencies, being some tens of megahertz, are much lower than the incoming off-air UHF.

The final block is the COFDM demodulator chip that takes the IF2 signal as input and delivers the MPEG2 (or MPEG4 in some other designs) digital transport stream as output to the remainder of the TV set. Facilities are provided for either the I/Q (in-phase/quadrature) data signals from satellite transmission or from the cable TV demodulator to control the digital processing chip.

5.3 Delivery systems: LNBs, feeders, and compact antennas

As shown in Figure 5.6, with terrestrial TV the UHF off-air input signal is generally taken directly from the antenna into the tuner and processed to form IF1, IF2, and then the MPEG2 digital processing. In most instances this provides for effective reception.

Digital TV differs from its analog counterpart in more ways than just the technology. The quality of the antenna, its reception characteristics, and its immediately local electromagnetic environment is of great importance. Reception critically depends upon the quality of the initial UHF reception at the antenna. High-quality antennas are recommended in order to guarantee excellent reception in all but the most forbidding environments.

In areas where the signal strength is relatively low a low-noise amplifier (LNA) can be interposed between the antenna and the tuner to boost the signal strength. Ideally the LNA should be positioned as near to the antenna as possible because only then is the noise figure maintained low. With digital transmission, noise is much less of a problem than with analog and therefore such an LNA could usefully be integrated within the antenna input side of the tuner. However, DTV subscribers are really in an "all or nothing" situation because they are either in the position of having exceptionally high-quality reception or they have absolutely no picture and most probably no sound either!

Low-noise blocks (LNBs), more precisely termed *low-noise block downconverters,* are required when the system is satellite rather than terrestrial. In this case the familiar dish reflector is the conventional form of receiving antenna and the LNB is almost always located at the focus of the dish. Many suppliers offer products with the dish, the LNB, and the final downlink coaxial cable feeder connection integrated within a single unit.

The satellite signal, at a Ku- or Ka-band microwave frequency, is picked up by the dish antenna and concentrated by this parabolic reflector, being focused upon the LNB input horn. The first stage in the LNB is an LNA which amplifies this signal. Further processing within the LNB then down-converts the signal into the first intermediate frequency. This is basically similar to the tuner function described above for the terrestrial DTV receiver.

More compact planar antennas implement technologies such as microstrip arrays. These approaches do away with the need for the bulky dish reflector and instead provide much more compact and environmentally friendly products. Since the gain (i.e., the focusing capability) of such compact arrays is generally substantially lower than the values possible with dish reflectors, it is desirable for the satellite transponders to have higher values of microwave output power. This is just about achievable with solid state semiconductor implementations rather than tubes such as TWTs.

Such planar antennas have a similar basic appearance to the layout shown in Figure 1.16, except that the MMICs are absent.

5.4 Technological developments and their likely impact

Many practically continuous advances in technology will impact the development of DTV during the third millennium.

Major drivers for these advances include such features as WebTV. With this type of system Internet services can be accessed and added to "normal" viewing, providing interactive services through an otherwise conventional TV set. Figure 5.7 indicates a product of this type, in this case manufactured by Pace Micro Technology (United Kingdom). The system is activated by the user's smart card and controlled from a matching and fairly conventional QWERTY keyboard.

Another possible driver is 3DTV. Advances are continually being made in this field and possibly the only significant drawback is the need

Figure 5.7   WebTV. (Courtesy of Pace Micro Technology.)

for the viewer to wear special glasses. If 3DTV takes off, then, in spite of bit rate compression techniques, it is highly probable that more bandwidth will be needed to carry the information-rich signal.

Effective digital signal processing (DSP chips) will also be demanded at ever-higher speeds. In 1999, for example, Texas Instruments (TI) launched a 160-MHz DSP chip designed for voice-over-IP, communications servers, and PBX add-ons. But many advanced DTV and associated applications require much higher speeds. The technologies surrounding "beyond 1-GHz" digital processing chips, already at the R&D stage at the end of the twentieth century, are proving vital to new third-millennium systems.

Another area of high significance concerns DTV VCRs and these use hard disk drives for storage as opposed to the more familiar videocassettes. Using the D-VHS (data-VHS) format, a signal input rate of 28.2 Mbps applies for the HS (high speed) mode as required for HDTV and

SDTV. Nontechnical issues including wrangles over copying and copyright protection have held back the introduction of commercial dedicated DTV VCRs.

Digital rewritable CDs are another matter. These need more than 9 GB of memory space in order to enable two hours of quality recording and replay. At the transition into the third millennium R&D laboratories were approaching 4.7-GB CD capability and further advances in both information density and access technologies will probably soon lead to greater than 9-GB availability.

It is important to appreciate that conventional cathode-ray tubes (CRTs), as used in most computer monitors and TVs, are inherently analog in terms of the manner in which the screens are addressed. In contrast most flat-panel displays are inherently digitally addressed. This applies to technologies like LCD and also plasma.

With several manufacturers either in full production or at least geared up for plasma screen manufacture, this looks like the most appropriate enabling technology. But do not expect the old analog CRTs to disappear like the dinosaurs. During the 1980s and 1990s many technology industry analysts had "their forecasting brains burned" by projecting just this type of scenario.

It serves us well to remember that old technology dies hard—very hard.

Manufacturing costs represent another important issue here and increased speeds or relatively revolutionary (in production terms) product approaches all tend to increase these costs.

As feature sizes decrease, for higher speed semiconductors, so manufacturing costs tend to spiral and a high level of confidence in the market is essential before committing to silicon (or other) fabrication ("fab") plant investment. The silicon fab disasters of the 1990s, notably for DRAM computer memory producers such as Fujitsu and Siemens, should never be forgotten and these scenarios must never be replicated in any other sector of the industry—including components for advanced DTV.

There are no easy solutions, but the industry must remain alert to and aware of all technological developments that may provide routes to reducing manufacturing costs.

In this context approaches could include:

- The selection of silicon-germanium (SiGe) or even gallium arsenide (GaAs) as the semiconductor rather than Si;

- Transmission line technology for the PCB cards (appropriate to higher speeds);

- Micro ball grid array (μBGA) for chip attachment rather than the surface mount approaches (SM) of the late twentieth century;

- Multichip modules (MCMs) wherever possible to effectively integrate groups of associated chips.

Doubtless, given the scale of the industry and its virtually unstoppable momentum, manufacturing problems will be overcome and new developments such as digital rewritable CDs and WebTV will be seen and used regularly within our homes and businesses.

Select bibliography

Booth, S.A., "Digital TV in the U.S.," *IEEE Spectrum*, March 1999, pp. 39–46.

"COFDM Terrestrial Digital TV Receiver," *Infinity*, Fujitsu Newsletter, Spring 1999, p. 5.

Pereira, F., "MPEG4: a new challenge for the representation of audio-visual information," *Picture Coding Symposium '96*, Melbourne, Australia, March 1996.

Yamada, O., H. Matsumura, and M. Sasaki, "Development of Satellite and Terrestrial Digital Broadcasting Systems in Japan," *IEEE MTT-S Symposium*, Baltimore, New York, May 1998, pp. 71–74.

6

CATV and SMATV Feeds

6.1 Streets and communities

Driven on by the ever-extending quality and quantity of available programs in many areas of life interest, CATV and SMATV continue to grow globally. CATV can be construed to stand for either cable or community television, but SMATV is universally understood to mean satellite master antenna television—sometimes termed *SDTV,* or satellite distributed TV, in the United States. At the dawn of the third millennium a conservative estimate puts the number of CATV subscribers in the main economically advanced nations at around 200 million—mainly in the United States, Europe, and Japan. When China and India are included this figure probably exceeds 300 million, and if "homes passed" are added the total of actual and potential subscribers must be around the one billion mark.

The average annual growth in numbers of CATV subscribers is indicated in Figure 6.1. From the millennium transition level of around 180

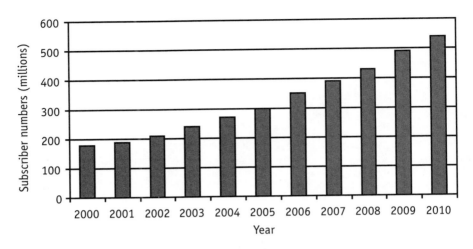

Figure 6.1   Global CATV growth 2000–2010.

million subscribers globally the number changes little at first due to the Y2K recessionary effect. Then after 2001 the inexorable rise resumes and total numbers are forecast to reach over 500 million by the year 2010.

Digital TV (DTV) and cable modems represent further important drivers for the expansion of CATV and SMATV. Cable modems permit relatively high-speed Internet access and these are therefore becoming highly popular.

Unlike off-air broadcast TV, satellite TV (DBS), CATV, and SMATV all have the advantages of enabling relatively large numbers of channels to be delivered to individual homes, apartment blocks, or hotels. Literally hundreds of programs can be delivered and the subscriber may use his or her remote control pad to request specific programs at the touch of just a few buttons. A wide range of types of programs are generally available, for example:

- Films—providing an almost limitless selection of movies;

- Sport of essentially all forms—but usually focused to the major local choices (e.g., basketball in the United States and soccer in the United Kingdom—although soccer has caught on internationally);

- News and current affairs programs such as CNN and Sky;

- Christian channels and other religions such as Muslim faith channels;

- Discovery Channel—in-depth coverage of features ranging from natural world issues to jet fighter technology;

- Cartoon and specialized channels designed for kids;

- Travel and shopping channels;

- Local news programs.

There are also many other offerings ranging from history to the salacious programs that are usually mistermed "adult."

The final item on the above list, local news programs, is of particular importance because in reality everyone lives first and foremost in their own local community. Any channel catering for the locality is therefore like an advanced up-to-the-minute version of one's local newspaper—often with live or recently recorded interviews and hearings. Such channels may cover just the local city or village, or they may stretch wider to include a localized region such as a small portion of a state or county.

CATV has a fairly long history. It all began with the extensive use of coaxial cable—well before the dawn of satellite communications and fiber optics but now both of these technologies form vital elements supporting the CATV industry. Today both fiber optics and satellite transmission are vital to most CATV networks.

Indeed many networks are hybrids of fiber and coax (H-F/C) with spectrum allocations: 5–30 MHz for voice, 54–550 MHz for analog video, 550–700 MHz for digital video, and finally 700–750 MHz for multimedia. Nellist and Gilbert's book shows these allocations in a spectrum allocation chart (reproduced as Figure 6.2 here) and also provides succinct descriptions of CATV including Internet access. Spectrum usage extends to 750 MHz, for voice multimedia.

It is highly probable that fiber will be used even more extensively in both new and replacement CATV networks, mainly stemming from the serious concerns associated with signal interference from copper cable networks. For example, in Germany evidence has grown showing that many conventional CATV networks interfere with air traffic

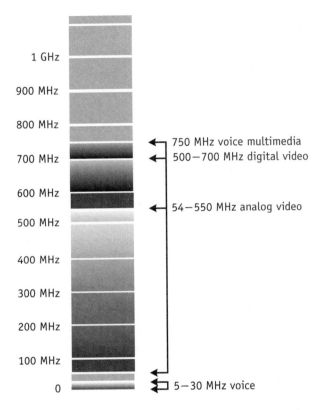

Figure 6.2   The proposed H-F/C (hybrid fiber/coax) spectrum allocation. (*From:* Nellist, J.G., and E.M. Gilbert, *Understanding Modern Telecommunications and the Information Superhighway*, Norwood, MA: Artech House, Inc., 1999.)

ground communication and navigation systems. Such serious concerns act as effective drivers for either highly screened domestic cable connections—or entirely fiber installations.

A typical CATV network begins with what is termed the *headend*. This may be fed with information from a wide variety of sources, for example, local studios, live outside/inside broadcasts, satellite downlink feed, locally or remotely stored film libraries, and the Internet itself. This headend, in most contemporary CATV networks, feeds the highly multiplexed main signal into a fiber ring via an array of laser transmitters and optical receivers.

In a simple configuration, as indicated for example in Figure 6.3, the optical signal is converted into an equivalent electrical signal within a local transition node. From this node the final connection is completed to the subscriber via coaxial cable to a cable modem which feeds both the TV set and also a personal computer (PC). An alternative is to have separate modems—one for DTV (set-top box) and one for Internet connection via the user's PC. However, in the latter case a separate subscription must be in place.

There are also usually provisions for telephone connections using the same CATV fabric, and this means that broadband video, telephony, and data (i.e., Internet or fax) can be carried by the same service provider—the cable company.

Of course in practice the multiplicity of functions required of a cable TV operator, the substantial distances demanded over a city area, and the often (desirably) large numbers of subscribers mean that a CATV network can never be as simple as Figure 6.3. Instead the more comprehensive network illustrated in Figure 6.4 is more representative.

Here ISP is Internet service provider, LEC is the local exchange carrier, IXC is international exchange, and PCS is personal communications

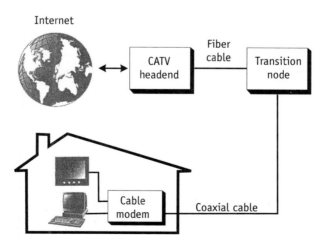

Figure 6.3 Internet access with CATV modem. (*From:* Nellist, J.G., and E.M. Gilbert, *Understanding Modern Telecommunications and the Information Superhighway*, Norwood, MA: Artech House, Inc., 1999.)

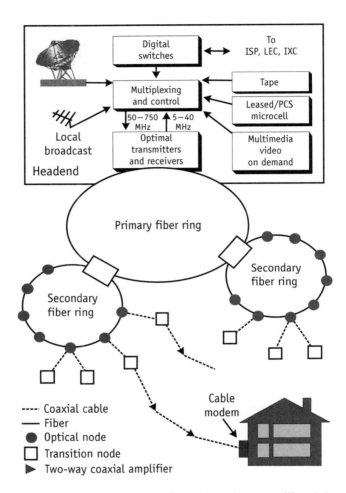

Figure 6.4 A typical CATV system configuration. (*From:* Nellist, J.G., and E.M. Gilbert, *Understanding Modern Telecommunications and the Information Superhighway*, Norwood, MA: Artech House, Inc., 1999.)

service—the term for mobile (noncellular) operators in the United States. In many other regions, Europe in particular, GSM or TETRA would be more applicable then PCS.

In these types of systems there is extensive deployment of optical fiber rings and links and the sharing of these. Only the final drops to the subscribers are coaxial and these are often limited to around 100m maximum from transition node to the home. In practice transition nodes take on a variety of appearances but small metal covered boxes near to hedges or street walls are fairly common.

The fiber rings and optical nodes usually follow street layouts, with occasional inspection and maintenance covers enabling access to the cable operator. For any unauthorized person to open and interfere with such a cover would be a criminal offense in the majority of instances.

Through the turn of the millennium the fibers were pervasively of the multimode type multiplexed at either 850 nm or 1,320 nm. These wavelengths are well below the optimum 1,550 nm that applies to the very low attenuation third window of single-mode fibers. However the bandwidths available with such multimode fibers are sufficient for CATV (and SMATV links), especially considering the relatively short distances involved.

Another important feature concerns the relatively large cross-sectional dimensions of multimode fibers compared with single mode. This makes transceiver module attachment and maintenance much easier—the system is more resistant to slight misalignments due to technicians failing to precisely fit components. This is extremely important because the most significant problem with CATV is service quality and outage.

In comparison, single-mode fibers are delicate and require skilled handling and fitting.

Notice (see Figure 6.4) how the initial off-air signal feed may be taken from either satellite or local (terrestrial) broadcast. Most of the features and facilities indicated in Figure 6.4 are commonly encountered in CATV systems within many developed economies.

6.2 Block dwellers and visitors

The types of CATV networks described above work well enough and provide good returns on investment when they apply to neighborhoods with many individual homes and small apartment blocks having perhaps up to around ten or so units. However, when large apartment blocks (e.g., with hundreds of units—usually termed *multidwelling units,* or MDUs) or hotels are considered, then it is usually not cost effective to take the signals from cable-company-owned and operated networks.

Instead the block or hotel owner will often be attracted to turn the operation into essentially its own CATV network. But in such cases, certainly with hotels, the concept of community hardly applies and another term is required to define this approach. Because many of these types of systems have traditionally been exclusively satellite-fed, with the antenna and LNB naturally rooftop-mounted, the name *satellite master antenna TV* (SMATV) was coined. An equivalent term, *satellite distributed TV* (SDTV), is frequently used in North America.

A schematic physical illustration of SMATV is given in Figure 6.5. The signal enters the satellite antenna—the "master antenna"—and is transferred to the head end via a substantial coaxial cable feed. From the head end the entire hotel block complex is fed with distributed TV channels. This distribution network may totally comprise coaxial cables

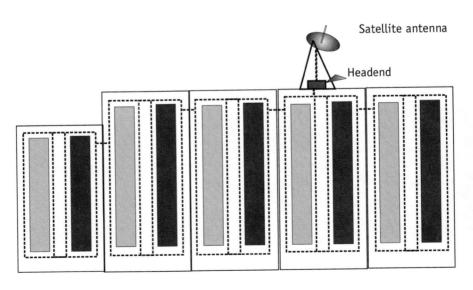

Coaxial or fiber cable downlead from antenna
and network interconnecting hotel rooms

Figure 6.5   SMATV in a typical hotel complex.

(common), or a mixture of coaxial and fiber cabling (increasingly), or perhaps occasionally entirely fiber.

Not shown in Figure 6.5 is the control center for the system that is usually located in the administrative staff offices of the hotel. Individual hotel rooms have the choice of many possible channels some of which will be "free" (i.e., combined with the price of the room), while programs on several other channels will require specific ordering.

Although often extending to several hundred meters, the distances to be spanned in large apartment blocks or hotels are generally much shorter than in city neighborhoods, so until recently fiber-optic transmission was rarely implemented in these installations and coaxial cable systems predominated. With the tangible advantages of freedom from electrical interference and capability for high-speed transmission, optical systems are becoming much more commonplace in MDUs at least. An example of a typical SDTV system topology is provided in Figure 6.6 in which solid curved interconnects represent coaxial cables and wavy line fills indicate optical fiber cables.

As usual, the signal first enters the satellite antenna and is then transferred along a coaxial cable downlead to the next unit. However, this first major unit is a fiber-optic transmitter capable of feeding from 8 to 16 optical output channels to possibly hundreds of nodes in the complex. The fiber-optic transmitter accepts the broadband electrical signal input and converts it into the multi-output series of optical signals. Serving each node there is an SDTV fiber-optic receiver that feeds each apartment with the required signal.

Within each apartment coaxial-cable-fed set-top boxes finally supply the signal to either the TV sets (or possibly WebTV sets) or to PCs having suitable modems, not shown in Figure 6.6. Such a system provides the advantages of fiber distribution in an economical manner.

Incidentally, the programming demands will differ markedly where hotels are concerned compared with the viewer profile associated with domestic units. Because large hotels generally cater to both national and international travelers, appropriate programs are required to suit these guests. Certainly local events, such as news and highly localized sports, will not be attractive to international hotel guests. These factors alone are powerful drivers for the hotel owners to introduce and run their own SMATV networks.

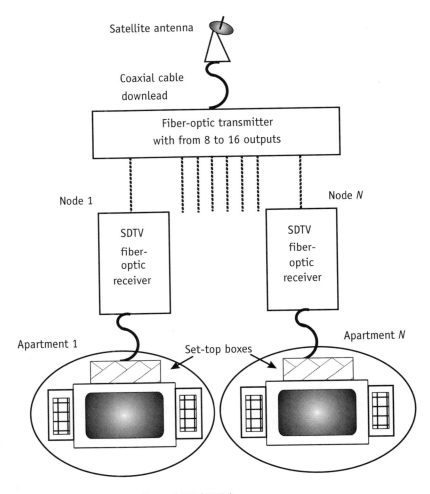

Figure 6.6   Satellite distributed TV (SDTV).

6.3 Community or "just you"

One key determinant in deciding upon the configuration of cable feeds is whether it is to serve a multitude of people or just an individual person or family.

For multichannel TV reception to individual home dwellers or those living in relatively small MDUs at least there is almost always the option of having individual satellite TV facilities. These traditionally provide receive-only access to (typically) a similar number of similar channels to those available from a CATV operator. Also, there is the independence from such an operator because it is easy to consider removing an existing installation and simply replacing it with a new one—perhaps obtained from a completely different store or mail order company. Even if the installation is not changed, there is always the opportunity of realigning the antenna to improve reception.

Basically with individual satellite TV facilities there is the perception (real or imagined) of much greater freedom and control over the incoming TV programs.

However, the prospect of having a vast population of satellite TV dish antennas littered across the outer walls of buildings such as MDUs is extremely unattractive and would almost certainly contravene local regulations and laws in many areas. Therefore, either the SMATV or the CATV approach is essential in these instances.

For individual or family units CATV can automatically ensure that with minimum effort the latest programs and program reception technology, such as DTV, are always available and that the subscribers are frequently reminded about this availability. With individual satellite TV, this is either leased through one of the large leasing companies or it is purchased outright. If leased, then the most up-to-date technology is immediately available from the owning concern so the situation is somewhat similar to that applying to CATV. If purchased direct, then the individual owners must pay for the cost of upgrades themselves.

6.4 Expansion capabilities

Issues have arisen concerning the drivers for DTV over CATV, and doubtless since most satellite TV is available in digital format this is rightly regarded as strong competition for the CATV operators. In fact these operators may be expected to provide some enhancements which others do not yet meet.

So CATV operators will increasingly offer digitally transmitted programs. A big question concerns the format and whether high resolution suited to HDTV is needed. Once this is decided, then a subtle but important technical question remains, namely how the QAM (quadrature amplitude modulation) used on several digital systems can be transferred through the 8-VSB format in the United States systems. One suggested technique for overcoming this is to remodulate the signal at the set-top box.

In 1999 at least two U.S. companies agreed to carry digital broadcasts: Cablevision Systems Corp. transmitting ATSC-format HDTV carrying major league baseball, basketball, football, and hockey; and Time Warner Communications Inc. transmitting CBS's digital broadcasts. Digital satellite TV transmissions are available via several operators, notably Astra, DirecTV, and United Motion.

By the late 1990s, notably in major European countries and North America, both CATV and SMATV still remained very largely receive-only. An important exception was the fact that many CATV operators offered telephony overlaid onto the system, usually beginning with conventional copper-wired final drops to the subscribers. This was followed by multiplexing through transition nodes onto the secondary fiber rings and then on up to the primary fiber ring. Another exception to the receive-only status was the facility for CATV subscribers to key-in orders for specific programs.

This approach suits narrow-band Internet access with modem rates up to 56 Kbps or somewhat higher. Faster access demands a rethink concerning the technology, and cable modems have provided one possible partial solution. However, a serious problem with shared cabling (especially with MDUs) is that the available bandwidth is strictly limited and when there are several simultaneous modem users the individual's speed of connection becomes greatly reduced. An interim approach toward solving this bottleneck is to install coaxial cables to the premises, but a much more far-reaching philosophy comprises installing fiber all the way into the final drop to the dwelling units or office blocks (all this with low-power electronics).

At least by the year 2020 there is a high probability that bandwidth demand will have reached the point where the multimode fiber rings that coped through the turn of the millennium are replaced by single-mode fibers. At that stage the likely scenario is single-mode primary and secondary rings and multimode final drops to subscribers. In this way the advantages of multimode fiber handling referred to earlier in this chapter are preserved where large numbers of drops of this type occur.

This introduction of single-mode fiber cables does not mean that highways and sidewalks (i.e., pedestrian pavements) have to be dug yet again for cabling provision. Which is good news because these "civils" typically take up anywhere from one-half to two-thirds of the total costs of a network installation—not to mention the secondary costs and inconvenience to everyone.

In most instances the earlier multimode cables and associated optoelectronics were installed in large-diameter loose tubes. This provided a high degree of "future-proofing," and physical access is straightforward with such installations. The new single-mode cables and their advanced photonic modules are overlaid within the same tubes and service inspection facilities.

Trends in fiber optics, satellite, and other transmission media are shown in Figure 6.7. In this chart the abbreviation *MW* refers to microwave and the abbreviation *MMW* means millimeter-wave.

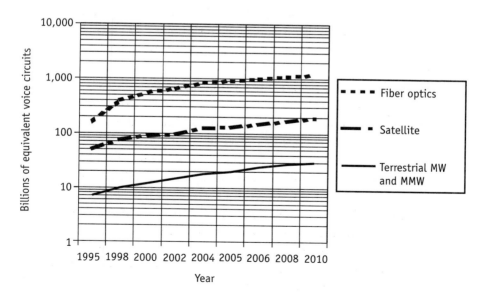

Figure 6.7 Global growth in transmission capacities. (From Peter Cochrane with updated data.)

It is important to observe that the vertical axis of this graph is logarithmic, ranging as it does from 1 billion equivalent voice circuits to 10,000 billion (i.e., 10 trillion) equivalent voice circuits. One equivalent voice circuit has a 64-Kbps channel capacity. Because of the necessary logarithmic scale the strong growth is masked to some extent. In fact the growth in all types of systems is very strong—but the actual capacities of fiber always remain relatively large, being about an order of magnitude greater than the nearest rival—satellite transmission.

Terrestrial microwave and millimeter-wave (MW and MMW in Figure 6.7) are becoming increasingly important and these, together with satellite systems, are subjects considered elsewhere in this book.

The provision of interactive services via satellite is much more problematic than with the CATV situation.

The main problem centers upon the sheer cost of owning and operating a satellite transceiver, rather than the relatively simple and inexpensive receive-only system for conventional satellite TV. With the familiar satellite TV outdoor unit all that is required is the dish (or planar) antenna with the LNB at its focus and the download. But when transmission is also demanded an RF (radio frequency—i.e., microwave) power transmitter must be installed and the download has to double as an upload. Furthermore, at the receiver the RF power must be isolated from the relatively sensitive receiver portion of the system thus complicating the arrangement.

The moderate-power transmitter unit can be located on the antenna feed, fairly close to the dish itself. This is a potentially hazardous unit from the emissions viewpoint and it also consumes substantial supply power. Most important of all, such transmitter units tend to be somewhat more expensive than LNAs, even when sold in substantial volumes.

The upshot of all this is that individual units (or small MDUs) can now expect to be able to afford two-way interactive satellite. Large MDUs and hotels could spread costs over substantial numbers of users and could therefore afford the capital cost of more sophisticated systems. The continued massive growth of Internet use will be a major driver for this development, but clearly it is the actual usage within specific MDUs or hotels which will determine whether such investments will actually happen.

6.5 Technological developments and likely impacts

In an overall sense, driven by the ever-increasing need for higher speeds of transmission (broadband Internet), the main trend in technology will almost certainly be towards more and more photonics. Current and near-future developments in fiber optics have been discussed earlier in this chapter.

In CATV the presently pervasive multimode fiber cables will be replaced by single-mode for the rings and multimode fibers will progressively be substituted for the copper drops to the subscribers. Of course this means that subscribers requiring the high-speed services will need to invest in new modems capable of interfacing with the multimode fibers.

This trend amounts to fiber-to-the-home (FTTH) and fiber-to-the-office (FTTO) as indicated in Figure 6.8. With both FTTH and FTTO the main feeder fiber cable runs along street-installed underground ducts—that is, the commonly encountered type of installation of the late twentieth century. In the case of FTTH the optical signal is taken from locally disposed feeder boxes that provide broadband optical signal feeds to each individual home. Because of the short ranges involved, these can be multimode fiber. Junction boxes in the homes make this signal available to the subscribers. The major driver here is people increasingly working from home, or a mixture of office and home interleaved.

FTTO is similar to FTTH as far as the street feeder box and indeed the multimode connection to the (now corporate) junction box. However, in the case of FTTO a more complex junction box is necessary because of the relatively extensive distribution of the signal around the corporation's premises. This distribution could also often involve fiber optics and can adopt a configuration not unlike that described above for CATV (Figure 6.6). In the corporate office instance satellite systems such as VSAT will also be involved.

It is reasonably safe to consider it unlikely that single-mode fiber will be demanded for the final drops to the home or to the office block for perhaps a generation at least.

Another important trend is toward the transmission of multimedia services by satellite, over broadband terrestrial (or stratospheric)

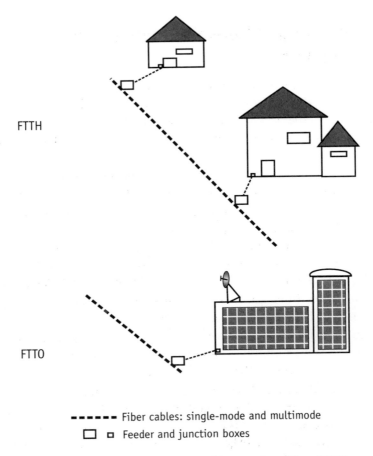

FTTH

FTTO

- - - - - - Fiber cables: single-mode and multimode
☐ ▫ Feeder and junction boxes

Figure 6.8 Fiber-to-the-home (FTTH) and fiber-to-the-office (FTTO).

wireless, and by fiber. Stratospheric and terrestrial options and development are described in Chapter 8 and broadband satellite is described in Chapter 9.

As always, the customer's requirements and then the provision of facilities form the most important drivers for all the technology. DTV and HDTV have already been covered above and also in Chapter 5. In the future, 3DTV may also prove to be a significant feature influencing technology choice in this sector.

Without any doubt the underlying technologies will be increasingly densely integrated in electronic, optoelectronic, and photonic forms.

Select bibliography

Booth, S.A., "Digital TV in the U.S.," *IEEE Spectrum*, March 1999, pp. 39–46.

BT Technology Journal, Vol. 16, No. 4, October 1998 (Special issue on access networks).

Clark, A., "Air Safety Prompts Cable TV Reviews," *Electronics Times,* London, England, April, 19, 1999, p. 1.

Cochrane, P., "The Future Symbiosis of Optical Fibre and Microwave Radio Systems," *Proc. European Microwave Conference*, September 1989, pp. 72–86.

Cochrane, P., Personal communications in 1999.

Dutta-Roy, A., "Bringing Home the Internet," *IEEE Spectrum*, March 1999, pp. 33–38.

Nellist, J.G., and E.M. Gilbert, *Understanding Modern Telecommunications and the Information Superhighway*, Norwood, MA: Artech House, Inc., 1999.

7

Teleports

7.1 What is a teleport?

In this chapter we first describe the nature of teleports and follow this up with a brief description of some examples. The various types of teleport are defined, again with examples of specific instances. A table is provided, comparing 13 teleports—most of which are subsequently described in considerable detail.

Where exclusively *corporate* high-speed Internet access is required, then very small aperture terminals (VSATs) adequately fit the need. There are over 386,000 VSAT sites worldwide, and the total is forecast to reach almost 800,000 by 2002.

Teleports are much more sophisticated than VSATs, being comprehensive, shared-use, and usually city-based telecommunications facilities. They are generally:

- Advanced;

- Intelligent;

- Capable of flexible, organic growth;

- Physically large.

The beginning of the third millennium marked the third decade of the teleport—a concept that began in the 1980s with small shared-use satellite communications "hubs" that were being developed by North American entrepreneurs. This essentially followed the deregulation of satellite services in North America.

Today's teleports include highly sophisticated developments, and the example of Technoport Osaka characterizes many of these. In this sophisticated facility the working-time population is reckoned to be between 80,000 and 100,000 people but for many teleports the numbers of incumbents is very much lower. The main "antenna farm" for Technoport is shown in Figure 7.1. Details are provided later in this chapter.

Naturally the explosive growth of the Internet has further spurred expansion in teleport activity because teleports provide good opportunities for broadband multimedia connectivity.

There are four basic forms of teleport and, working upward hierarchically, these are indicated conceptually in Figure 7.2. Each of these is now described, working upward from the lowest in the hierarchy.

The telecom-port form is characterized as follows:

- A telecommunications hub providing services on a profit basis;

- It may be either independent or carrier-owned;

- An "information port" rather than an airport or a seaport;

- A facility providing for the shared use of complex and expensive facilities.

This is really the most basic form of teleport, that is, a port for telecommunications. The basis is identical to that of any port including airports or seaports and amounts to the attractive economics of sharing the costs of a relatively comprehensive yet expensive facility.

Figure 7.1 The Osaka Media Port (OMP) in Japan. (Courtesy of Osaka Media Port Corporation.)

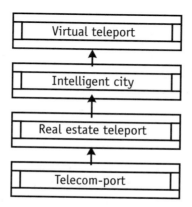

Figure 7.2 The four basic forms of teleport.

Telecom-ports may be independent operations or they may be subsidiaries of major carriers or even other corporate parents. They usually combine both satellite and terrestrial (increasingly fiber-optic) communications and they provide practically all the communication needs of any conceivable customer. This includes the most complex and bandwidth-hungry services imaginable and full motion videoconferencing is almost always provided.

Several Japanese teleports are telecom-ports: Osaka, Minato Mirai, and Tokyo being prime examples. New York's Staten Island teleport and the London Teleport are also telecom-ports. Most are described later here.

The real estate teleport has the following general characteristics:

- It adds intelligent and fully networked commercial real estate to the telecom-port;

- Usually developed by a public/private partnership;

- Provides value-added services to its tenants including upgraded land upon which these tenants can build and expand their operations;

- There is an overall appeal to multinational companies.

The real estate teleport is an extension of the fundamental telecom-port which is extended by means of an intelligent fully-cabled network that is often a mix of coaxial cable and fiber optics (see also CATV)—increasingly mainly fiber. This approach opens up new economic development opportunities as well as provides a secondary profit center, and the local region in which this type of teleport is developed can promote it as an ideal facility for companies with extensive telecommunications needs—and that is most companies.

Given this raison d'être an alliance between (local) government and business continues to be the major pattern for the development of most real estate teleports. Clearly the businesses look for the profitable opportunities while the government agencies interests are principally those of engendering local economic development. These real estate teleports are often particularly attractive for second- or even occasionally third-world economies because they provide world-class office and telecommunications facilities that help to pull in large multinational corporations.

The main Korean and Malaysian teleports are all real estate types. Other examples are the Cedar Hill Texas Gateway in the United States and Sunderland Teleport in the United Kingdom.

The intelligent city is the next level up in terms of sophistication and this is illustrated conceptually in Figure 7.3.

Clearly visible are microwave and millimeter-wave antennas for both satellite and terrestrial communications. Some of these provide base station/switching center interconnects for mobile (cellular) communications while others are dedicated terrestrial communications links. Not visible, because they are deeply embedded, there are also extensive upgradable fiber-optic networks and networked computer systems.

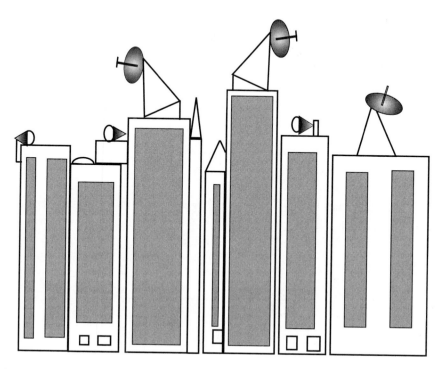

Figure 7.3 The intelligent city.

In essence this is either an entirely new urban center or the intensively redeveloped downtown region of a city. There are state-of-the-art communications and information services made available to business, academe, government agencies, and other institutions both on-site and also off-site. All the systems are linked to a central teleport operations center. Because advanced systems are already in place, occupying companies can simply "plug-in" their network components, and few if any enhancements are necessary.

This type of teleport is becoming extensively implemented in many economies, predominantly in Japan, some European cities (including Paris), and also the Rio de Janeiro teleport.

The virtual teleport is at the top level of the hierarchy shown in Figure 7.2 and it is also the most recent in terms of concept. This type of teleport virtually integrates existing advanced telecommunications infrastructures usually with some added value from introduced subsystems and services. Virtual teleports are:

- A cooperative venture among business corporations, academe, government agencies, and nonprofit organizations;

- An asset for infrastructure-rich communities who have no requirement for new telecommunications facilities or high-tech business parks;

- A one-stop shopping gateway to existing networks including satellite communications and associated services.

Those communities that are already rich in terms of their telecommunications infrastructures have no need for yet another new high-speed system. This is also an important consideration when markets are somewhat depressed, as during the moderate year 2000 Y2K-effect induced recession.

The virtual teleport provides for a pooling of resources between the types of organizations mentioned above and creates a virtual telecom and information hub by integrating assets owned by such diverse organizations. Although no formal, physical, teleport exists, individual and corporate users view it through their phones and multimedia computer terminals as though they were using a teleport.

The key point is that to establish a virtual teleport no new telecommunications infrastructure is necessarily required. Instead, a teleport is evolved from the system of intelligent interconnects available from the existing infrastructure.

It is almost certain that virtual teleports will steadily gain ground as the third millennium progresses—mainly in already highly developed economies such as Korea, Japan, Singapore, the United States, and parts of Europe. Cyberjaya, described in Section 7.3, is an example of an existing virtual teleport because otherwise disparate facilities are virtually interconnected on demand. The existing telecommunications infrastructure forms the basis of this teleport.

Across the entire teleport scene, for all types of installation, major international corporations have substantial stakes and these include, for example, British Telecom (BT), Cable & Wireless, SouthWest Bell (United States), and NTT in Japan.

7.2 Roles of teleports

Much has already been described in Section 7.1 concerning the roles of teleports and these roles include:

- Providing a highly efficient on-ramp for the Internet and the information superhighway;

- Providing services on a profit basis;

- Providing for the shared use of complex and otherwise expensive facilities;

- Appealing to multinational corporations as well as to other types of users;

- Providing for a pooling of resources.

Teleports should not be seen as separate stand-alone entities but rather as installations or even virtual concepts that become naturally integrated with other surrounding and international systems and networks. This applies to essentially all forms of communications systems, including CATV and SMATV or SDTV. The concept also applies to primarily data networks such as LANs, metropolitan area networks (MANs), and wide area networks (WANs), and examples of the possible interconnections are shown in Figure 7.4.

Connections from SMATV or CATV networks are predominantly one-way whether between each other, between them and the teleport, or from CATV to MAN. All other interconnections, to and from LAN, MAN, or WAN and teleport, are two-way real time and most are optical fiber. MAN interconnections are shown at 155.52 Mbps (SDH level-1), corporate WANs operate at kilobits per second rates, and the standard 64-Kbps rate is indicated. A corporate LAN increasingly has gigabit

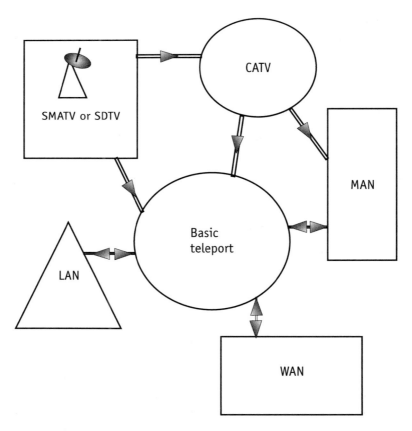

Figure 7.4 Teleport with CATV, SMATV, LAN, MAN, and WAN interconnections.

Ethernet backbone capabilities and therefore a 1-Gbps information transfer rate is shown here.

Teleports are highly flexible and can be transparent to the types of technology involved. In terms of services, it is of no consequence whether cabled or wireless interconnections are involved. The specific type of technology applied at various nodes will be appropriate to the geography and topology existing and the information throughput required. Another very significant aspect is the capability for organic growth, since old systems or subsystems may be replaced by new and more appropriate subsystems as the requirements change and need upgrading.

Most important of all, once "future-proofed" networks are in place, the emphasis can shift toward software implementations of as many upgrades as possible. In short, as teleports are software managed and controlled, they can be software reconfigured and upgraded.

In the future a major trend will probably be for the physical growth of cities and other similar communities to slow, but such entities will expand constantly in terms of information and communications growth. It is in this respect that teleports will have increasingly influential roles to play.

7.3 Teleports of the late twentieth century— and beyond

In this section examples of current teleports and also some of those in development are described.

These types of large-scale developments are all essentially under continuous development and what is presented here is the situation at the time of writing together with anticipated developments where some indication of these can be interpreted. A summary of several teleports, distributed globally, is provided in Table 7.1.

There are probably over 100 teleport projects truly globally distributed and their numbers are frequently being added to. It is realistic to say *global* because such projects are located in most countries and not only the G8 or "first world" economies.

Table 7.1

Selected Teleports: Types, Owners, Operators, and Users

City and Name	Country	Type	Owners/Operators	Users
Rio de Janeiro	Brazil	IC	Mainly public investment	BI
Osaka Teleport	Japan	TP	Osaka Media Port Corporation	DI Including SkyPerfecTV and VPN
Tokyo Teleport	Japan	IC	Public investment and Japanese broadcasting	BI (mainly)
Minato Mirai 21	Japan	IC	—	BI (mainly)
Pusan	Korea	RE	Pusan City Government and SK Telecom	BI
Jong Jung	Korea	RE	Korea Telecom	BI
Cyberjaya (Putrajaya): MSC	Malaysia	V	Telekom Malaysia, R&D organizations, airport authorities	BG
Singapore (ST Teleport)	Singapore	IC	Singapore Technologies	BG
Cedar Hill, Texas Gateway	U.S.	RE	SouthWest Micronet	DI
New York, Staten Island	U.S.	TP	Port Authority and GlobeCast N. America	DI
London, Kingston TLI	U.K.	TP	Kingston Communications	DI
Sunderland Teleport (Doxford Business Park)	U.K.	RE	Sunderland City Council (with Fnet as substantial operator)	BI—also broadcasting, education, and training
Padua (Padova)	Italy	IC	Interporto Consortium and the Local Industrial Park	BI—also local government and education

Legend: MSC = multimedia supercorridor

Types:
IC = intelligent city
TP = telecom port
RE = real estate
V = virtual

Users:
BI = business and industry
BG = business and government
DI = domestic and industry
(In all cases almost always including a large proportion of Internet traffic.)

Although doubtless still characterized by continuing future growth at various rates, teleport projects in several developing economies have been influenced by the local economic difficulties experienced in the late-1990s. In particular this applies to Batam in Indonesia, Korea, Malaysia, Pakistan, and Thailand. The Korean projects are in Seoul and Pusan, while Malaysia has three projects: Cyberjaya, IMT-GT, and Labuan. Pakistan and Thailand each have one teleport project—in Karachi and the Eastern Seaboard respectively.

7.3.1 Japanese teleports

In Japan there are at least seven major teleport projects: Tokyo, Osaka, Yokohama, Sendai, Nagoya, Fukuoka, and Kobe. The largest and most mature of the Japanese projects are the Tokyo Teleport, the Osaka Teleport, and, 25 miles southwest of Tokyo, the Minato Mirai 21 Teleport in Yokohama.

Many projects are of the telecom-port or real estate teleport classification, but several would be classified as intelligent cities.

The Tokyo Teleport occupies 448 acres and is distributed between two artificial islands in Tokyo Bay. An automated light rail system serves the teleport and a fiber-optic WAN is integrated within the facility, encased in underground utility conduits that form a spider's web of interconnections. Everything is controlled from a high-tech building called the Tokyo Teleport Center that is topped out with satellite and terrestrial microwave and millimeter-wave antennas. Associated structures include a convention center, intelligent offices, and the headquarters of a major Japanese broadcasting company. A total of U.S.$3 billion in public funds has been invested into this project with at least as much in private investment. Although the project had almost bankrupted the city by the end of the twentieth century, it is still slated for completion by the year 2003.

The Minato Mirai 21 Teleport creates a new city environment, based upon information and communications technology. Information-based services support business and local communities.

Well to the southwest of Tokyo, the Osaka Teleport—or Technoport Osaka as it is also known—is located on the shores of Osaka Bay and represents a truly vast and comprehensive telecommunications port. The Teleport Center was completed in 1998 and houses all the communications and power equipment.

The frequencies used for the satellite communications associated with the Osaka Teleport include C-band (6/4 GHz), Ku-band (14/12 GHz), and Ka-band (30/20 GHz), and close carrier frequencies are also used for several other services in Japan (e.g., 14/12 GHz for VSATs). It was appreciated that if precautions were not taken with the site, interference would have been a problem. Therefore an interference-free environment was created around the main antenna "farm" taking the form of an eight-meter high bank of earth topped with evergreen trees. A view of the two main antennas located here is shown in Figure 7.1.

By using this environment Ku-band interference has been reduced by 99% and for both the C- and Ka-bands no interference occurs. From Figure 7.1 it can be seen that the antennas have a somewhat unusual geometry since there are twelve-sided dish reflectors. This reduces the extent of undesired sidelobes of microwave field distribution.

An organization known as Osaka Media Port (OMP—hence the logo on the dishes) is responsible for the running of services from this teleport where it employs 631 people. OMP also owns approximately 23,000 km of fiber-optic cabling distributed around the Kansai region and signals at 150 Mbps can be transported over this network.

OMP was established in 1985 but its Osaka teleport operations really began in 1996 with what is now named the SKY PerfecTV! Service—a markedly successful DBS service covering much of Japan. The first Internet connection service offering was in 1997 (WCN, or World Computer Network) and an ATM-leased circuit service was started in 1998. Toward the end of 1998, OMP began offering the Internet VPN (virtual private network) service, termed OMP SafeNet.

A 30-Mbps satellite digital transmission service is currently operational with this teleport. Around the year 2000, or shortly thereafter, OMP may be upgraded enabling high-speed on-ramp access to the information superhighway.

Outside of Japan several further teleport projects are in place or under development.

7.3.2 The Rio de Janeiro Teleport Center

The Teleport Center in Rio de Janeiro, Brazil, involves U.S.$40 million of public investment and is a highly advanced project which includes at least 23 "intelligent buildings." This U.S.$40 million investment may appear low in comparison with several other projects in more advanced economies, but it is substantial for a country like Brazil, and there will also be considerable private investing.

This project began in 1997 and is due for completion by the end of year 2000. There is a massive (400,000 square meter) infrastructure, and the Teleport Center itself occupies 160,000 square meters of space. At the top of this center there is room for all the required satellite antennas as well as helicopter pads. A 60,000-line Teleport Operations Center has also recently been completed.

Telecommunications capabilities include:

- Services from C- to Ku-band;

- A virtual network server;

- National and international satellite up/down links;

- Conversion facilities for all broadcast systems;

- High-speed coding/decoding facilities;

- Local and remote video exhibition and recording;

- Transponder purchase, rental, and management;

- Digital compression and expansion;

- IDR (intermediate data rate: 64 Kbps to 44.736 Mbps) services;

- Full-motion videoconferencing and business TV;

- Electronic and voice mail, individual and programmable mailboxes.

The Rio de Janeiro Teleport appears set for substantial future development and only a considerable downturn in the economy would arrest or delay such advances.

7.3.3 Teleports in Malaysia and Indonesia

Malaysia has two teleport projects, the largest of which is the Multimedia Supercorridor (MSC). This project is an important part of the Vision 2020 on which the country is embarked. The economic downturn that began in 1997 slowed this and many other Pacific Rim developments, but third millennium recovery will enable expansion to become reestablished.

MSC is a high-tech zone interconnecting three centers: Cyberjaya—a 1.6 million square meter research and development campus, a new administrative capital city taking up 15,000 hectares in Putrajaya (the new capital city), and the extensive new international airport of Malaysia. A gigabit digital fiber-optic network is in place to serve this extensive MSC—applications for which include:

- Borderless marketing centers;

- Electronic government;

- Telemedicine;

- Value-added services;

- Worldwide manufacturing networks.

There is also strong software development in this project because Microsoft is a player.

Telekom Malaysia plays a vital part in this project and customer-driven communications is considered as the principal aspect. Microsoft Corporation is also involved in this development, and this is especially interesting because Microsoft is also expressing a keen interest in the trans-global Cable & Wireless Company. The indications are that the well-known software corporation is strongly entering the advanced telecommunications scene.

Another project, private-sector funded and known as the IMT Teleport, is jointly operated by the Indonesia-Malaysia-Thailand Growth Triangle Project on LangKawi Island. A fundamental problem here is that all three countries involved experienced the 1990s economic downturn referred to above, and as a result this project has developed only slowly.

In any event, the LangKawi development is substantially short of being a true teleport so far.

Both Indonesia and Thailand have their own nascent teleport developments that came under economic strain during the late 1990s. The site of Indonesia's development is Batam Island and the first phase of development here was completed in 1997. Pasifik Satelit Nusantara (PSN) has introduced a satellite ground station to serve the substantial number of international companies that have factories in the large industrial park and resort center on Batam Island.

The Eastern Seaboard Teleport (EST) of Thailand functions as an international communications gateway located 130 km from the capital, Bangkok. Operated by TOT and CAT, EST serves the Lemchaban industrial zone and the two operators interconnect their satellite center with a fiber-optic WAN.

7.3.4 The intelligent island—Singapore

In the final quarter of the twentieth century Singapore became known universally as "the intelligent island." This country has long been in the forefront of high technology with relatively early implementations of intelligent networks. Specifically, for many years extensive IT networks have been operational and there is a broadband nationwide public-switched fiber system.

Singapore Technologies Pte. Limited was set up in 1994 to spearhead the government-controlled "ST" Group's investments in telecommunications services in Singapore and the region. The avowed aim was to become a leader in the telecommunications field by drawing on the multi-disciplinary competencies and synergies of the complete group, as well as collaborating with other leading players who possess complementary skills and experience.

There was a planned entry into telecommunications, CATV, and multimedia services that enabled Singapore Technologies to establish a strong presence in Singapore, China, the Philippines, and elsewhere in the Asia-Pacific area.

ST Teleport was set up in 1994 as an integrated full services provider for the broadcast industry. Its $25 million state-of-the-art satellite and broadcast facility has been designed to offer a full range of comprehensive services under one roof. With its strategic location and dish antenna farm that enables extensive connectivity, ST Teleport has the ability to access regional and international satellites such as Apstar, PanAmSat, INTELSAT, and AsiaSat. To fully utilize this connectivity, the company has also taken on ad hoc transmissions for news bureaus and multinational corporations (MNCs).

In 1997, ST Teleport added a new $1.6 million production studio, including video facilities, to meet expanding broadcast needs. The addition of this new facility enables ST Teleport to cater to the needs of production houses and MNCs as well as advertising agencies. The fully equipped 150 square meter production studio has been the venue for commercial shoots, variety shows, and "live" business television. ST Teleport, with its complete range of high-quality facilities, is ideal for MNCs, government bodies, and medical associations to hold events such as "live" product launches, training programs, and globalized distance learning.

Singapore Telecom operates three teleports, principally for access to INTELSAT and Inmarsat and also to control the ST-2 satellite in which this organization has an investment jointly with the Taiwanese.

Singapore inaugurated its Teletech Park Building in the Science Park Building II in 1996. With services operated by Singapore Technologies this facility serves as a hub supporting the development of innovative network services to businesses, research centers, and universities.

The ST Teleport is shown in Figure 7.5.

7.3.5 South Korea teleports

In Korea two "teletropolises" are planned. SK Telecom is joining with the Pusan city government and private companies to develop a $2.5 billion redevelopment of the harbor district to form a teletropolis. This is planned to integrate exhibition, financial, seaport, and trade facilities.

The second Korean teleport development is led by Korea Telecom and is located 50 km to the west of Seoul on Jong Jung island. Although not officially named, this teletropolis might be termed Jong Jung Teleport. It services an offshore industrial and finance center with advanced telecommunications.

Figure 7.5 The ST Teleport in Singapore. (Courtesy of ST Teleport.)

In a similar vein to the problems experienced by most other parts of the Pacific Rim in the late 1990s, South Korea also had its own economic problems, and these have to some extent moderated the development of both teletropolises.

China is also developing advanced teleport-like facilities, notably in Beijing and Shanghai, and in Pakistan the Karachi Teleport is under development—with a special emphasis on attracting software companies.

The teleports in Hong Kong are also highly significant. An INTELSAT earth station antenna associated with one such teleport is shown in Figure 7.6 and Hong Kong Telecom's facilities are shown in Figure 7.7.

7.3.6 Teleports in the United States and Canada
There are many teleports located in the United States, and most integrate CATV and related programming into their systems. A historic example is

Figure 7.6   INTELSAT earth station antennas in Hong Kong. (Courtesy of Bruce Elbert, *Introduction to Satellite Communication, Second Edition,* and Hong Kong Telecom.)

Figure 7.7   Hong Kong Telecom's Teleport Facilities. (Courtesy of Bruce Elbert, *Introduction to Satellite Communication, Second Edition,* and Hong Kong Telecom.)

that of The Teleport in New York City (Staten Island) which was opened as long ago as 1985. This remains a good example of economic regeneration driven by this type of investment, and in this case $70 million was spent by the Port Authority to develop the required infrastructure.

The fiber-optic network associated with this Staten Island Teleport extends into adjoining areas of Manhattan and Brooklyn. Merrill Lynch has a worldwide data center at this location and a new company, Telehouse, is responsible for operating the mainframe computers and telephone switches here. The satellite center is owned and operated by GlobeCast North America.

An interesting group of U.S. teleports are those provided by South-West Micronet. These implement terrestrial interconnects, satellite services, and fiber-optic networks to provide all of the types of services mentioned under other teleports described above. SouthWest Micronet's Texas Gateway is on a 27-acre site at Cedar Hill in Texas where there is an international fixed earth station. This company also operates a satellite facility in Dallas.

Another teleport project is located in the Niagara Falls area of Southern Ontario (Canada) and the Buffalo region of New York. This will be termed the Niagara Teleport and it will be financed by a combination of public (the Empire State Development Corporation) and private money. The Buffalo region already has an advanced fiber-optic network and this joint effort is expected to accelerate local investment and to bring new jobs to the region.

Another North American facility, the PanAmSat teleport in Fillmore (California), is shown in Figure 7.8.

7.3.7 U.K. teleports and related operations

Across the United Kingdom several teleports are in operation.

As the major U.K. telecommunications services company, British Telecom (BT) is naturally prominent. For example, BT Broadcast Services operates four major earth stations in the United Kingdom, at Goonhilly, Madley, London Teleport, and Martlesham Teleport. Each teleport, connected by terrestrial fiber to the BT Tower in central London, plays an integral role in the company's global transmission network. BT uplinks over 70 full-time TV services via its main earth stations,

Figure 7.8 The PanAmSat Teleport in Fillmore, CA. (Courtesy of Hughes Electronics Corp.)

for major broadcasters including Canal Plus, Multichoice Africa, BBC Worldwide, Modern Times Group, MTV/Viacom, Flextech, BSkyB, and others.

Outside the United Kingdom, BT Broadcast Services owns and operates teleports in the United States, France, and Moscow. All three state-of-the-art facilities were designed and built by the company over the 1996–1999 period.

Located in the heart of the U.S. capital, BT Broadcast Services' Washington D.C. Teleport is staffed by a team of industry professionals. The facility offers end-to-end connectivity between BT and its partner teleports globally through its Ku- and C-band satellite transmission routes. MPEG-2 4:2:2 compression equipment installed at the Washington, D.C., facility station enables variable bit rate and high color definition for

BT's clients that require back-hauling of U.S. channels into Europe, Asia, or Latin America.

BT Broadcast Services provides what are termed *on-site up-links* for the design and operation of teleports directly at the broadcaster's premises. Many of the company's clients prefer dedicated on-site up-links for essentially three reasons: cost, simplicity, and peace of mind.

- *Cost:* Often on-site up-link systems are more cost effective because the terrestrial loop to an independent teleport is eliminated. Additional play-out fees can also be taken in-house at a significant cost saving to the broadcaster.

- *Simplicity:* An on-site system can be simpler because it relieves the broadcaster of the requirement to deliver programming to the teleport and to manage the external relationship in the event of a fault.

- *Peace of mind:* All of the BT Broadcast Services on-site up-links have dedicated redundancy and uninterruptable power supply systems. Services operated through the on-site systems are also remotely controlled and monitored, 24 hours a day, by qualified staff at the London Teleport. Broadcasters can feel confident that they are receiving the appropriate high quality of service on their own premises.

The BT Tower in central London is the United Kingdom's broadcasting hub for networked transmissions. Over 1,000 program switches are made each day sending programs between broadcasters and production houses around the United Kingdom and globally, through the tower's direct connection to and from all BT teleports.

Another significant U.K. teleport operation is at Chalfont Grove, west of London, and in 1998 this was acquired by the U.K. company Kingston Communications (KC). This teleport is the satellite, TV, and broadcast up-linking facility for Services Sound and Vision Corporation (SSVC, a U.K.-based charity). As part of this development, KC's satellite business, Kingston Satellite Services (KSS), entered into an outsourcing contract to supply SSVC, and its British Forces Broadcasting Service, with all their worldwide requirements for satellite services, radio and TV studios, transmission design broadcast systems, and TV facilities.

KSS is one of the United Kingdom's leading satellite service providers, with a substantial portfolio of "blue chip" clients. KSS's products include fixed up-link and down-link services for both analog and digital TV applications, as well as for audio, teletext, and high and low speed data. It provides Internet delivery to end users and Internet service providers (ISPs), as well as forward-and-store services, all based on its low-cost DVB platform.

KSS is also subcontracted by Racal Telecom to operate the hub for the U.K. National Lottery satellite network—one of the biggest VSAT networks outside North America. Other clients include News International and the BBC World Service.

The teleport itself is operated by Kingston TLI located at Chalfont Grove, approximately one-and-a-half hours to the west of London, and a photograph of this facility is provided as Figure 7.9.

A further major teleport project is the one in Sunderland—at the Doxford International Business Park. This represents a multimillion dollar investment that turned Sunderland, around three hours north of the capital by rail, into Britain's first teleport outside London.

This project was implemented by an organization known as TCI Corporation and the teleport offers advanced telecommunications access to about 85% of the world's population. Direct access is provided to the Orion F1 digital satellite, covering both Europe and North America. The following services are provided:

- Broadcasting;
- Educational and training facilities;
- Large-scale data transmission;
- Multimedia design;
- Private corporation networks;
- Videoconferencing.

In 1995 Sunderland City Council recognized the role an earth station satellite link teleport could play in adding value to the developments

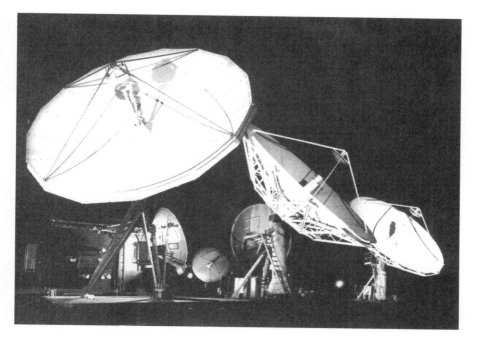

Figure 7.9 Kingston TLI's Chalfont Grove Teleport in the United Kingdom. (Courtesy of Kingston TLI Ltd.)

underway in the city. These were intended to help expand the economic base away from traditional industries such as shipbuilding, coal mining, and heavy engineering. A teleport would help to underpin the move towards a more knowledge-based and IT-based service sector. From the beginning, the city council recognized that the development of a teleport would be a key opportunity to link business development to broader social and regeneration aims, not just in the city, but across the northern region.

The council sought out a private sector telecommunications partner (TCI, a London-based technology and communications company) for the development of the earth station satellite link up facility. The council then worked with its private sector development partner for the Doxford International Business Park site (Akeler Developments) to secure the location for the teleport. It then facilitated a range of other partners in the city and region to come together to occupy a building (the teleport) to provide a focal point for telematics-based services in the region. It subsequently played a role in the formation of an appropriate telecommunications corporation (Northern Gateway Telco, or NGT).

The award-winning teleport is sited in Sunderland in the northeast of England. It comprises a building of 14,000 square feet offering high specification serviced accommodation. In addition to the satellite receive and send up-link facility operated from the building, the teleport houses a range of partner organizations engaged in telematics-based activities, and a photograph of this facility is provided as Figure 7.10.

Also on the 50-acre Doxford International Business Park and the adjacent 20-acre high specification Doxford Technology Park are a range of blue chip companies. These include the European headquarters (HQ) of Nike, the HQ of Arriva (the biggest public transport operator), One to One (mobile phones), London Electricity (sales and services), Royal Sun Alliance (financial services), the HQ for Northern Rock (Building Society

Figure 7.10   The City of Sunderland Teleport (United Kingdom). (Courtesy of Leighton Internet.)

services), Subscription Services Ltd. (the commercial arm of the United Kingdom's Post Office), AVCO Trust (financial services), Camelot (National Lottery), and Barclaycall (a major center for online banking). Over 6,500 people are currently employed on the Doxford Site.

Through the Loral Orion F1 Satellite the City of Sunderland Teleport offers satellite communication access to Europe and the western seaboard of America. The scope goes as far west as Denver, as far north as Oslo, as far east as Kiev, and south to Algiers in North Africa.

Eventually, when the Loral Orion F3 Satellite is launched and safely in orbit, the coverage will be extended to the Pacific Rim countries and the Far East, with an international gateway in Hawaii.

The launch of the Loral Orion F2 satellite, on October 19, 1999, extended the footprint to cover the midwest of the United States, east of the Urals, South America, South Africa, and the Middle East. This brought 85% of global population within the range of the Teleport. This will be supported by Loral Orion's global gateway network of six digital video broadcast platforms, located in six major cities around world.

A 2.4m dish currently carries a Worldcast E Burst 256-Kbps duplex circuit (with 4 Mbps burstable on return path) connecting the Sunderland Host (the city's ISP) to the U.S. Internet backbone on the eastern seaboard of the United States (Mae East). The specification is subject to imminent changes as further new services are launched, with the Loral Orion solution being fully scalable.

The City of Sunderland Teleport is a strategic regional telecommunications gateway to the north of England, introducing advanced satellite and Internet voice capabilities to the region's knowledge businesses and offering a route to the global marketplace for northeastern content. The Teleport is operated by Northern Gateway Telco. This is a joint venture between TCI and local partners—the University of Sunderland, Northern Informatics, Leighton Internet, and Multimedia Ltd.

The City of Sunderland Teleport was given the "Developing Teleport of the Year Award 1996" and "World Teleport in Development Status" in 1998, by the World Teleport Association, based in New York.

In April 1999, the role of the teleport and the partners' approach to laying the Foundations of Excellence for regional networking was recognized by the jury of the Global Bangemann Challenge. The city has been listed as a finalist in the New Business Structures section of this global competition, which seeks to find the best uses of ICTs to promote economic and social well-being.

Through the Loral Orion satellites, the Teleport offers a full range of competitive international leased line services including point-to-point and point-to-multipoint able to offer voice, data, video, videoconferencing, and fax as well as Internet capability. Internet services include the Worldcast and Worldcast E Burst brands (ISPs). This teleport's current capacity of half a million minutes per month is scalable to the demand. The teleport offers co-location and caching services to ISPs and is host to the Northern Interconnect, a network of companies within the northeast of England.

Specialist services, including *Dynalink*, provide additional functionality for customers with dynamic requirements. In 1999 a unique approach to year 2000 problems was being put forward by avoiding vulnerable nodes on a global terrestrial network and offering disaster recovery and capability, which can form part of a flexible and long term sustainable solution. At the time of writing, the full effectiveness of the Y2K approach remains to be seen.

The teleport is also host to Internet Voice capacity, provided by U.S.-based FNet Corporation, over PVC (private virtual circuit) circuits.

Sunderland City Council's vision for the future is to ensure that the city stays ahead in the race for investment and jobs. This underlies the commitment to put the city at the leading edge of the telecommunications revolution and the new world of opportunity this offers.

With the launch of the region's first teleport, Sunderland is rapidly becoming the United Kingdom's most advanced telecommunications center outside London. As of the year 2000, three satellites linked the city to 90% of the world, 24 hours a day, promising major benefits to the city's industries.

Proposals to extend the City of Sunderland's Teleport include creating a Global Teleport Training Center which would become the key U.K. training establishment for all matters relating to teleports. A center to assist new telematics business start ups is also to be considered, providing carefully tailored support, business advice, accommodation, and training.

Businesses are not the only enterprises to benefit from the teleport. Thousands of people across the region will enjoy greater opportunities for training and employment, involvement in city decision-making, and cheaper access to the Internet and international phone calls. The aim of the Sunderland City Council is to enable every person, in every part of the community, to become involved in the new information society. The city's first Electronic Village Hall, a community-based telematics center, has opened providing local residents with opportunities for training, tele-working, and communicating with special interest groups, and, by the end of 1999, all primary and secondary schools will be linked up to the Internet.

Other proposed community initiatives include:

- The creation of an Intelligent Housing Estate, providing electronic access and online services from each householder's home;

- Cyberskills Workshops, to help the unemployed gain IT skills needed by potential employers;

- The development of the award-winning city library's computer information access facility into an IT training venue.

This U.K. city is now recognized locally, nationally, and internationally as a world center for telematics.

Sunderland Teleport also has a gateway via FNet Corp that provides access to an international network of voice and fax services that typically provide 40% to 50% cost savings for businesses. A global training center is available at Doxford International Business Park and approximately 2,000 new jobs are expected to be created as we transition into the third millennium.

7.3.8 Italy

Following an extensive consultancy study by the World Teleport Association (WTA, headquartered in New York), the project for Italy's first teleport—Teleport Padua—was started. Padua (or Padova as it is also known) is located about 20 miles west of Venice, in the northeast of the country. This teleport is being constructed on a 1,200 hectare (3,000 acre) industrial zone owned by an Industrial Park and Interporto Consortium. The following goals were established by the consultancy mission:

- To link the Veneto Region with a new infrastructure tool and to combine the strengths of each city in the region, hence creating a domestic and international marketing advantage.

- To support and expand the small- and medium-size business users (the SMEs) currently doing business inside the industrial zone of Padua.

- To create a "beehive" or magnetic effect in order to attract ancilliary business to the teleport footprint.

- To link the regional, municipal, and national administrations (i.e., public sector), and also the universities, research facilities, Info-Camere, and other major users, providers, and "repackagers" of data services. This will turn Padua into an intelligent city.

- To ensure the stability of the business base by establishing a teleport with a forward-looking, innovative system that will monitor the ever-changing telecom and technology needs of the user communities and to apply appropriate methods to adapt to the required changes.

Dr. Amedeo Levorato heads this development. At the time of writing in mid-1999, Teleport Padua is under advanced development, and therefore Italy's teleport progression would appear assured well into the third millennium.

7.4 Likely future scenarios for teleports

How a particular teleport develops, and even what form of teleport is envisioned for a specific region, depends critically upon the pattern of user needs and the local economic circumstances at the time of conception and during the early-build phase.

As described in Section 7.1 it is possible to define four basic forms: the telecom-port, the real-estate teleport, the intelligent city, and finally the most advanced concept of all—the virtual teleport. Well into the third millennium it is entirely probable that the world will continue to see examples of each of these basic forms being established in various locations.

Further, it is important to appreciate that existing teleport projects will continue evolving so that projects that began as telecom-ports or real-estate teleports will become essentially intelligent cities.

Virtual teleports are different—in that they virtually integrate existing advanced telecommunications infrastructures—usually with some added value from introduced subsystems and services. Cities ripe for this form of development are always those that already have a sophisticated information infrastructure but not formally a teleport. Suitable cities could include Austin (Texas), Birmingham (United Kingdom), Buenos Aires (Argentina), Capetown (South Africa), Delhi (India), Denver (Colorado), Dubai (United Arab Emirates), Guangzhau (China, PRC), Hong Kong (China, PRC), Leeds (United Kingdom), Leningrad (Russia), Los Angeles (California), Madrid (Spain), San Diego (California), Singapore, Sydney (Australia), Stockholm (Sweden), Tel Aviv (Israel), Vancouver (Canada), and many more cities where the communications infrastructure is already advanced.

As the third millennium develops momentum, many cities in the G8 first world economies will gain intelligent city projects. In North America examples could include Detroit and New Orleans, while in Europe cities such as Berlin (being the new seat of government and with much advanced industry), Dresden, and Glasgow (United Kingdom) could benefit.

In the second- and third-world nations the situation is naturally different and more complex. India, for example, has a strong legacy in satellite communications and advanced industry, and yet its city centers need substantial redevelopment. Major cities such as Delhi and Calcutta could become candidates for telecom-ports or real-estate teleport projects. In regions such as Central and South America, apart from Brazil where the Rio de Janeiro Teleport is developing, there is almost no history of advanced industry and any new projects such as telecom-ports may be delayed many years.

Global operators are likely to emerge, to offer teleport project development and management at any level suited to the local situation and budget. This may possibly be started by an organization like the World Teleport Association that has considerable consultancy experience in this field.

The concept of a "teleport enterprise corporation" (Global TEC) may become relevant here. Such an organization would provide a highly flexible service such that it could efficiently and dynamically manage practically any form of teleport project in the world. An impression of this concept is given in Figure 7.11.

In this speculated schematic four operating projects are indicated, one in each of four countries: India, South America, Germany, and the United States. It must be emphasized that this is purely speculative and that as far as is known, no such projects actually exist at present. The three "nebulous bubbles" are intended to indicate potential projects that have not so far reached the signed-up stage. All operating and potential projects involve two-way highly interactive dialogue as shown by the arrows.

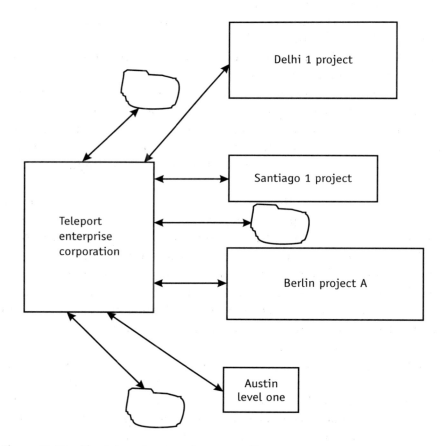

Figure 7.11 The teleport enterprise corporation.

This dialogue would be efficiently run via telecoms as far as possible with occasional physical visits, only where absolutely necessary.

Most likely the main focus should be on common language connections in which those nations that share a (largely) common language may have preferential channels between teleports in various countries. Australia, New Zealand, North America, and the United Kingdom would feature strongly in this respect with English being the working language. Chinese would also be increasingly important, mainly covering the vast expanse of China, and Spanish would be an added language of great significance throughout Central and South America as well as Spain itself.

As projects develop to sufficient levels of maturity so the teleports themselves would be used to increasing extents as on-ramps for super-highway communications.

7.5 Teleportation technologies

As described in the foregoing sections, teleports inherently involve all or almost all forms of technology: satellite, terrestrial radio (microwave and millimeter-wave), fiber optic, copper coaxial cables, and computer (both hardware and software). Few, if any, other types of telecommunications installation include this all-embracing family of technologies.

In common, however, with many systems it is only the larger sub-units that are visible on a day-to-day basis to the casual observer—and this mainly comprises the satellite and terrestrial antennas. Staff operating within a teleport complex of whatever form will, of course, be all too familiar with computer screens full of information every day.

Behind all this lie generally invisible but extremely important entities, and these include the electronic cards, the fiber optics and other cabling (usually contained in ducting), and the software housed in disk memory on the computer systems.

An interesting numerical comparison can be made by taking the ratios of typical physical dimensions relating to specific parts implemented within such comprehensive systems.

At one extreme, the antennas may on occasion have reflector diameters of around four meters (just over 12 ft). On the other hand, a great many system components have devices with microscopic dimensions. Examples include the internal core of a single-mode optical fiber that is around eight microns in diameter and the feature sizes applying to the fastest transistors in the integrated circuits on the electronic cards that can be as small as 0.15 microns.

Applicable ratios therefore are:

$$\frac{(\text{Antenna diameter})}{(\text{Fiber core diameter})} = \frac{4}{(8\times10^{-6})} = 500{,}000$$

$$\frac{(\text{Antenna diameter})}{(\text{Smallest transistor feature size})} = \frac{4}{(0.15\times10^{-6})} \approx 26 \text{ million}$$

In the physical technologies implemented in teleports it is therefore clear that enormous dimensional differences apply. Although this is generally true of much communications electronics, it is particularly marked in this case.

On the microwave, millimeter-wave, and satellite side, most suppliers are based in North America. This also applies to the computer hardware and software products required for teleports. However, fiber-optic cables, photonic components, and related products are available from a range of companies globally.

Millimeter-wave technology is growing in importance, and increasingly, therefore, more corporations are found that can supply products capable of functioning at frequencies above 30 GHz.

A general trend for teleports and other related telecom installations is toward more millimeter-wave and an intensification of photonic and fiber-optic content.

In the longer term, perhaps around the second decade of the third millennium, everyone's home also becomes a teleport. Some of the possibilities surrounding this concept are mentioned in Chapter 10.

Select bibliography

Chan, Ted, from ST Teleport Pte. Ltd., e-mail: ted@sttp.st.com.sg.

Elbert, B., *Introduction to Satellite Communication, Second Edition,* Norwood, MA: Artech House, Inc., 1999.

Kingston TLI at http://www.kingston-tli.co.uk.

Levorato, Amedeo, from Teleport Padua, e-mail: info@xxi.it.

Osaka Media Port Corporation Corporate Profile and other information (Nakanoshima Center Building, 19th Floor, 6-2-27 Nakanoshima, Kita-ku, Osaka 530-6691, Japan, e-mail: teleport@po.omp.co.jp).

Sunderland Teleport (UK) privately obtained information.

World Teleport Association Updates (regular seasonally updated information) www.worldteleport.org.

8

Terrestrial and Stratospheric Broadband Wireless

8.1 Broadband wireless without satellites or fiber

Most of the high-profile advances in broadband and high bit rate communications, increasingly for Internet on-ramps, are the province of new and planned satellite constellations and DWDM fiber systems. The latter are described in Chapter 3 and broadband satellite systems are discussed in Chapter 9.

However, neither satellite nor fiber represents the only possibilities for the delivery of broadband wireless services—far from it. It is well known that microwave or millimeter-wave signals will travel through the atmosphere, and in fact this is how it all began.

During the second half of the twentieth century the term *microwave* generally meant one of two possibilities: either line-of-sight terrestrial communications or else something one cooked with very efficiently. Before fiber optics gained the undisputed lead in terrestrial communications systems, line-of-sight microwave links with their dish-festooned towers were clearly visible across most industrial countries. Prior to the 1980s most links were analog, using FM and FDM, although more recently digital transmission has been implemented.

Many of these types of systems, now almost entirely digital, remain in operation, and in some instances the installation of new links continues. This approach is vital in regions where it would be completely uneconomic to install fiber-optic cables due to either low population density or rigorous terrain, or both these factors. These types of links often involve very broad overall bandwidths, to accommodate large numbers of channels, but the individual services are usually of a relatively narrowband nature.

In this chapter we are concerned exclusively with broadband systems and the types of links referred to above are not considered further here.

Why is it that neither satellite systems nor fiber-optic transmission can be expected to deliver all the broadband services required, well into the third millennium? This is an especially interesting question because, ultimately, most communications traffic and indeed most information processing will almost certainly be by photonic means, with wireless taking up the mobile needs locally. This scenario, however, is still many years away and we deal with some of these developments and issues in Chapter 10.

So fiber optics is still insufficiently mature to economically handle most broadband traffic, even in metropolitan areas. Wireless provides an excellent solution for the last mile (e.g., local access). In which case, will not satellite systems always prove to be the decided winner for customers living and working in regions of low population density or difficult terrain?

Well, the planned broadband satellite constellations are getting more and more advanced, as described in Chapter 9, but there are problems inherent with such high-cost projects. All of the planned systems operating at bit rates above 30 Mbps, except for the V-band systems, involve total system investments exceeding $2 billion, and Teledesic is expected to cost at least $9 billion.

The V-band projects that may come on-stream around 2010 or somewhat later could cost even more.

The challenge here is of a basic financial business nature—namely the rate of return on the investments (or ROI as it is often termed). In 1999 the Iridium narrow-band LEO project was already showing the signs of a slower-than-anticipated uptake and this could be a foretaste of things to come.

Another feature applicable to both fiber optics and satellite systems is the time-to-deployment. Entirely new fiber-optic cabling projects take years to plan and install. However, as mentioned in Chapter 3, where existing fiber cabling installations already have narrow-band multimode fibers, these can be fairly readily upgraded for the broadband future at dramatically reduced costs to the provider. Eventually, as existing cables age, this will become a further tremendous advantage for cabled networks.

Satellite launches also take at least several months to complete and in the meantime launch failures are an all-too-frequent occurrence. Apart from the obvious financial losses and precious lost time, the additional expense in terms of increasing insurance premiums also has to be taken into account. Yet another factor is the immense difficulty and cost of any maintenance—usually the prospect of astronauts seriously servicing communications spacecraft is more than daunting!

On the other hand, it is important to note that broadband satellite will largely be suitable for mobile users and this is a vital advantage.

Even given the strong growth in mobile users with their demands for on-the-move communications, there will continue to be large numbers of customers who need broadband telecommunications, notably for Internet access, from their homes and workplaces. This requirement is driven onward by the fact that these homes and workplaces are increasingly identical locations.

Observing the above challenges for both fiber cabling and satellite communications, several substantial industry players have, so to speak, brought things closer down toward the ground, and by the end of the twentieth century strong growth had been established in terrestrial multipoint/multimedia systems. Some organizations have initiated fascinating and innovative programs in which either very high flying aircraft or even stratospheric balloons are beginning to play significant roles as communications transponder platforms.

The take-up for all types of broadband services, current and forecast, is shown in Figure 8.1. This includes all forms of systems, not just terrestrial and stratospheric, but the latter are definitely anticipating substantial and growing slices of the action.

These types of systems are described here.

8.2 Terrestrial options

A total of four types of microwave and millimeter-wave distribution systems are entering service to varying extents in many economies and these are summarized in Table 8.1.

The first point to note from this table is that while MMDS operates within the relatively low frequency band of 2.1–2.7 GHz, all the other options are at substantially higher frequencies and have much greater

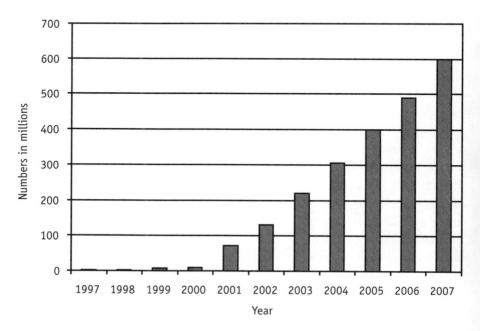

Figure 8.1   Global subscribers to all types of broadband services.

Table 8.1

Microwave and Millimeter-Wave Distribution Systems

System	Acronym	Service Frequency Extremities (GHz)
Multipoint microwave distribution system	MMDS	2.1–2.7
Local multipoint distribution services	LMDS	27.5–40
Microwave video distribution system	MVDS	10–43
Multimedia wireless system	MWS	10–43.5

bandwidths. The actual available bandwidth for MMDS is only 202 MHz. This service is useful for situations where the less directional microwave beams assist in terms of setup and multiple users of single beams, provided the relatively low bandwidth is acceptable. Nellist and Gilbert cover MMDS well in their book and this is not considered further here.

The remaining three systems, LMDS, MVDS, and MWS, offer truly broadband services and therefore substantial attention is paid to each of these in separate sections here. However, it is important to note that the frequency extremities shown in Table 8.1 are the extremes over which specific services may be selected for operation and they are not the band edges of specific systems. The actual operating bandwidths are typically less than 1 GHz or sometimes as high as 1.4 GHz.

MMDS is less demanding of line-of-sight due to use of the relatively low 2-GHz band. LMDS, and all the other types of systems, require LOS in all cases due to the higher frequencies. This results from the increased directivity at these higher frequencies (see Chapter 1). Above 20 GHz more frequency reuse is available—which multiplies the effective bandwidth even further.

A comparison of bandwidths for various services is shown in Figure 8.2 and this has been developed from Figure 2.6 (lower diagram) in Chapter 2. Although the 202-MHz bandwidth for MMDS is fairly wide, it is clear that the total available bandwidth for the other services is much wider—being a factor of five greater.

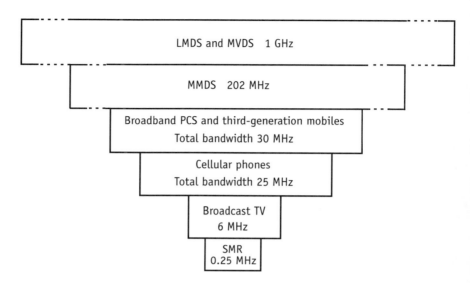

Figure 8.2   Bandwidth for services including terrestrial broadband wireless.

At these millimeter-wave frequencies path precipitation such as rain or snow significantly attenuates the signal, a phenomenon known as "rain fade." During such periods the transmitter output power is increased automatically in order to compensate for the otherwise reduced signal strength. However, the combined effects of precipitation and terrain can drive a need to increase equipment deployment, hence leading to increased investment.

LMDS is now the most important of all these systems and services. This technology is gaining considerable acceptance in many countries and looks set for strong growth.

We consider LMDS in more detail in the next section.

8.3 Local multipoint distribution services (LMDS)

As remarked in Section 8.2 above, LMDS is a system of great importance (global market forecast at $3.5 billion by 2003), and therefore it is described in some detail here.

The overall system concept of LMDS is shown in Figure 8.3. While PC terminals are indicated here in practice, customers tend to have a mix of PCs and TV sets with digital set-top boxes and, increasingly, TV Internet adapters.

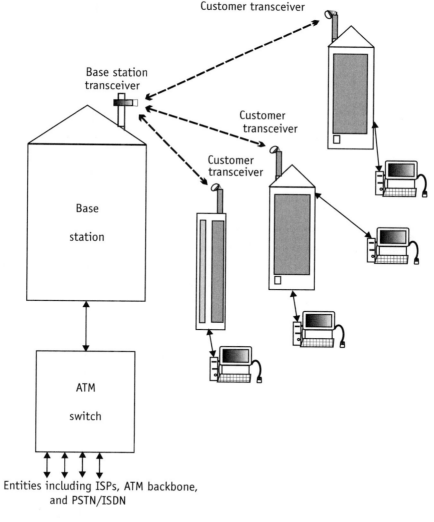

Figure 8.3   Overall concept of LMDS.

Each customer has a transceiver capable of direct interactive LMDS operation and these transceivers intercommunicate with a base station transceiver using frequency channels in the locally prescribed LMDS range. The base station is connected via an ATM switch to external networks, for example ISPs, the ATM backbone, and PSTN or ISDN.

Over the 1997–1998 period several system topologies were developed by various systems integrators. Typically, STM-1 (SDH) connections were facilitated between the ATM switches and the base stations.

A more comprehensive and state-of-the-art type of network is indicated in Figure 8.4. The ATM switch, base station, and transceiver configurations are essential elements for this network that is now highly LAN-, multidwelling unit-, and commercially oriented. On the cabled network side of this system, fiber optics is used to connect with the base station on OC-3, at 155.52 Mbps.

With LMDS in general, transmission on individual links may occur at speeds ranging from 40 Mbps to 20 Gbps.

Spectrum allocations for LMDS in the United States are shown in Table 8.2.

In the United States, Art Radio Telecom Corp., Teligent, and WinStar Communications are substantial players, although Canada was originally in the vanguard with this technology because in this country early licenses were granted for the 27.5–28.35-GHz band. In Canada, MaxLink Communications of Ottawa, for example, has a license to provide telecommunications services in 33 cities.

Cisco Systems, Marconi Electronic Systems, Motorola, Newbridge Networks, Nortel Networks, and SpectraPoint are all examples of important suppliers to these markets internationally.

The specific benefits of LMDS for broadband access are the relatively:

- Rapid deployment;

- Fast ROI and service revenue onset;

- Low network maintenance, management, and operating costs.

Rapid ROI and service revenue onset is expected because of the demand for Internet access at higher speeds. This approach is also particularly attractive to SMEs that cannot afford fiber installation.

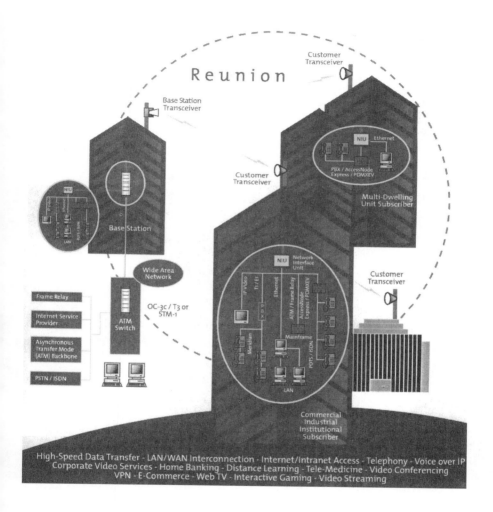

Figure 8.4   The "reunion" network for broadband wireless access. (Courtesy Nortel Networks Broadband Wireless Access [BWA].)

With this approach the operating costs tend to shift away from the infrastructure and toward the customer premises equipment (CPE).

Another important advantage is that these systems are designed on a "grow as you go" basis, meaning that none of the system installer's capital becomes wasted when customers churn, as they always do.

Table 8.2

Spectrum Allocated for LMDS in the United States

Block	Effective Bandwidths (MHz)[1]	Frequency Ranges (GHz)
Block A	1,150	27.5–28.35
	1,150	29.1–29.25
	1,150	31.075–31.225
Block B	150	31.0–31.075
	150	31.225–31.3

[1] Following frequency reuse. (There is also a 38.6–40.0-GHz licensed band.)

TDMA or FDMA access techniques are alternatively applicable, although different pros and cons are associated with these according to the application environment. LMDS networks use ATM radio interface cards (ARICs) to accept and modulate the ATM traffic, translate it into baseband IF, and transmit it over conventional coaxial cables to rooftop transceivers. These ARICs may use either TDMA or FDMA. With TDMA there is a shared bandwidth environment that is a useful facility for a large number of subscribers. On the other hand, FDMA enables a high-connectivity environment to be established that is useful in multiple-dwelling units, for example.

To date it would appear that no commercial operator has opted for CDMA although this technique may well be considered in the future.

Newbridge Networks' MainStreetXpress multimedia LMDS system considerably enhances the basic topology. It uses FDMA and enables data to be transferred at rates ranging from 800 Mbps to 12.8 Gbps. One example of a Newbridge Networks installation is that of the Home Telephone Company of Charleston (South Carolina) where fast Internet services, wireless, paging, data, and also POTS are all being offered on a single system.

In contrast to Newbridge, Stanford Telecom offers ARIC modems functioning on TDM and employing TDMA for customer access. Also, in 1998 Lucent Technologies acquired the previous Hewlett-Packard LMDS business at the same time establishing its own Wireless Broadband Division. Scalable hardware is the watchword for Lucent in this context, and their GlobeView 2000 broadband switch will operate at bit rates from 5 to 20 Gbps.

Meanwhile WinStar Communications of New York, referred to above, claims that in the top 50 U.S. markets there existed (in 1999) around 750,000 office buildings with less than 10% fiber connectivity available and only growing slowly. Since it costs about fifteen times as much to lay fiber in such buildings, compared with only $20,000 to facilitate LMDS, this would appear to be the winning approach until broadband fiber really comes into its own in one or two decades. Advanced Radio Telecom Corp (ART) has joined forces with Lucent Technologies to perform a similar approach.

Examples of operators in Europe include Swisscom in Switzerland where the previous MVDS (see Section 8.4) is being extended to enable commercial subscribers access using digital services and also Telia in Sweden that has LMDS deployments.

In Japan the Kobe earthquake destroyed much of the fixed and mobile infrastructure in that region and prompted the authorities as well as businesses to seek an alternative to fixed line infrastructures. This led to a serious consideration of LMDS, and its adoption is now fairly widespread in both Japan and South Korea.

8.4 Microwave video distribution and multimedia wireless systems (MVDS and MWS)

Predating LMDS and originally conceived and deployed for analog TV (video) transmission, microwave video distribution systems (MVDS) necessarily became increasingly digital and essentially multimedia.

It might therefore be considered that MVDS is now just another name for LMDS, but there are subtle distinctions. For a start, the TV connection remains preeminent in the architecture and an MVDS set-top box decoder is generally included. Also the operating frequency ranges tend to be different from LMDS because for MVDS these can be as low as 10 GHz or as high as 43.5 GHz. The use of these bands enables MVDS to be deployed in regions where licenses can be obtained for these frequencies rather than the LMDS 27.5-GHz to 31.3-GHz range. Another feature is that rarely if ever with MVDS is the term *base station* used. Instead the term *headend* tends to be employed, akin to CATV concepts.

The basic architecture of MVDS is shown in Figure 8.5 in which the MVDS link is indicated with bands that may be anywhere within the 10-MHz to 43.5-GHz range.

The system is ultimately connected to a node on the SDH network via another broadband link to the control, data, and voice transceivers that provide the forward and reverse paths via a rooftop-mounted antenna. This entire assembly is the headend. In practice there could be several similar MVDS links interconnecting other transceivers on further buildings and eventually the information is displayed on the customer's TV or PC.

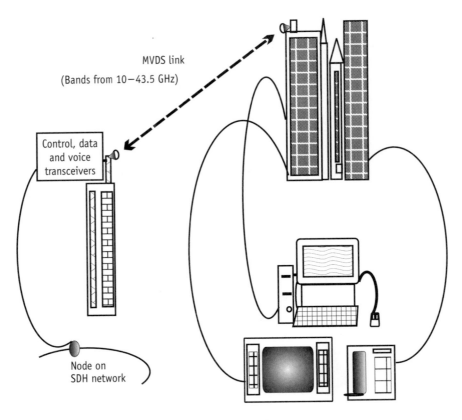

Figure 8.5 Overall conceptual layout of MVDS.

Control and monitoring is performed at the lowest bit rate in the system—9.6 Kbps.

A 40-GHz MVDS transmitter is shown in Figure 8.6 in which the two radome-protected linear antennas and the vertical cooling fins are clearly visible.

Frequency reuse, sectorized directional antennas, FM, and both horizontal and vertical polarization are commonly used. Cells in the

Figure 8.6   A 40-GHz MVDS transmitter. (Courtesy of GEC-Marconi.)

network are typically between 4 and 5 km in diameter. Dynamic bandwidth allocation with TDMA is generally employed and this provides spectrum on demand by the customer. Techniques for digital modulation include QPSK, 16-QAM, and probably in the future trellis-coded modulation techniques.

Allocations for frequency spectrum were provided at the World Radio Conference in 1997 (WRC '97) and for Region 1, covering Europe mainly, the 40.5- to 43.5-GHz band was allocated for MVDS. Other frequencies such as 10.5 GHz have also been allocated for the provision of integrated services using wireless local loop (WLL). This is becoming increasingly important for underdeveloped and developing economies, and it is probable that Kosovo, for example, could well be the recipient of these types of services.

Countries such as Hong Kong and Qatar have opted for 12-GHz MVDS and at least one Qatar system uses analog rather than digital transmission. A 12-GHz, 35-cm diameter receiving dish, and low-noise block (LNB), as installed in Qatar, are shown in Figure 8.7.

Most of Europe, including Germany, Sweden, Switzerland, and the United Kingdom, operates within an allocated 40.5- to 42.5-GHz band. Unit prices for receiver LNBs have been falling toward and down through the $150 mark. Transmit unit prices, however, remain stubbornly high as a direct result of the relative difficulty of obtaining MMICs that will deliver sufficiently high power at these high frequencies.

MWS systems are very similar to the MVDS systems described above. One exception is that the upper frequency limit in the allocated range is, at 43.5 GHz, somewhat higher than for MVDS.

Like LMDS, access techniques are either FDMA or TDMA and the same comments apply here regarding CDMA to those mentioned under the LMDS heading (see Section 8.3).

8.5 Platforms using stratospheric balloons and high-flying aircraft

There are space and satellite communications, discussed in Chapter 9, and then there are the terrestrial systems and services described in Section 8.4 above. However, very much closer to ground level than

Figure 8.7 A 12-GHz MVDS receiver with 35-cm dish antenna and LNB (installed in Qatar). (Courtesy of GEC-Marconi.)

LEOs we can have balloons and, of course, aircraft. Some enterprising people have seriously considered the possibilities of using balloons, strictly "aerostats," or aircraft as transponder platforms for broadband communications services.

Until recently the available technology associated with balloons essentially ruled these out as serious contenders in this role. Aircraft did not appear promising either for several reasons—notably the fact that most planes spend much of their time on the ground at airports where their use for communications systems is virtually precluded. These types of platforms have been used quite extensively by the military both now and in the past.

Interestingly, the station-keeping technologies developed for commercial satellites, including the extensive use of satellite navigation (e.g., GPS), means that both balloons and high-flying airplanes can be closely controlled thus paving the way for their use in communications platform roles.

8.5.1 Balloons (aerostats)

Various players are actively interested in deploying stratospheric balloons as platforms for broadband services. Probably the best known is Sky Station International who intend to fly their first commercial balloon for this purpose in year 2000. Sky Station's ultimate plan is to have a fleet of about 250 balloons in service, hovering at 21-km altitude above many major cities, at some time before 2005. Each balloon can carry a payload up to a maximum of around 1,000 kg and power is obtained from a combination of fuel cells and solar arrays.

The 21-km altitude can readily be imagined, being just over twice the typical altitude (marginally below 10 km) at which we are all familiar when flying in commercial jets. By 2005 these fleets of balloons and other craft will probably become familiar sights for aircrews, their passengers, and indeed many people on the ground.

A substantial industrial consortium comprising Aerospatiale, Dornier Satellitensysteme, and the Jet Propulsion Laboratory (owned by NASA) is in place to pursue this program. Technological problems that are yet to be solved include the difficulties of operating in the extremely cold environment—it is not often appreciated that the upper stratosphere is actually colder than outer space because of conduction and convection. There are also the not inconsiderable difficulties involved in maintenance and the replenishment of consumables.

Another airship-based network has been proposed by the Japanese Ministry of Posts and Telecommunications, and Turin Polytechnic (Italy) is cooperating with the Italian Space Agency (ISA) to develop and construct a long-endurance aerostat powered by fuel cells for night operation and solar energy by day. At 100 kg the payload capability is only 10% of Sky Station's and the available electrical power is also relatively low. But this Turin/ISA project could still be very much a winner given the anticipated advances in practically all electronic hardware, including microwave and millimeter-wave solid state device (MMIC) capabilities and their ever-decreasing supply power requirements.

Also, the Turin/ISA program is expected to only involve a $3 million investment and to cost $345 per hour to run. For a start, the $3 million is at least an order of magnitude lower than LMDS or MVDS and around two orders of magnitude below the investment levels associated with most broadband satellite projects.

Yet another project, known as RotoStar, is being developed at Tel-Aviv University in Israel. Like both Sky Station and Turin/ISA, RotoStar is designed to hover at the 21-km altitude. Unlike any of the other schemes publicized to date, however, the platform vehicle for this project would remain on station for periods between four days and six months. A critical distinction with RotoStar is the fact that it is being designed as an unmanned rotating-wing craft rather than strictly a balloon.

8.5.2 High-flying aircraft

The main contender in this respect is known generically as HALO—for High Altitude Long Operation—and here this means a fleet of high-flying aircraft each carrying broadband transponders.

Since 1996 the Raytheon Company had been working with another corporation known as Angel Technologies to demonstrate the feasibility of providing broadband wireless services using HALO aircraft. By late summer 1998 the two players entered into a formal teaming agreement for this program and demonstrated the basic feasibility of the concept.

Angel Technologies' HALO aircraft design, called HALO-Proteus, can carry a large antenna high above the center of a city to deliver very high-speed communications services. Augmenting terrestrial microwave towers and orbiting satellites, the HALO aircraft will fly fixed-pattern trajectories in the stratosphere in order to provide metropolitan wireless services at a lower cost, with increased flexibility and improved quality of service, than either satellite or fiber. A variety of fixed and mobile wireless services are offered including voice, data, and video.

In August 1998 a test flight was conducted using ground and airborne electronics developed by Raytheon in which the companies jointly demonstrated a record-breaking 52-Mbps wireless link between the ground and Angels' HALO Network testbed aircraft in flight.

Under the teaming agreement, Raytheon will have responsibility as the prime electronics systems integrator for both the airborne and ground segments of the HALO network. This company will leverage its extensive capabilities in complex integrated RF systems and digital technologies to architect the HALO network.

By late 1998 Angel Technologies announced an agreement with Wyman-Gordon Company for the certification and production of 100 HALO-Proteus Aircraft, and in summer 1999 this aircraft made its debut at the high-profile Paris Air Show. Part of a fleet of HALO-Proteus aircraft, flying at less than 90 knots above cities, is shown in Figure 8.8.

The platforms comprise relays of HALO-Proteus aircraft circling for periods up to 16 hours at 16 km altitude, providing footprints on the ground (*cones of commerce* is the term used here) that can have diameters of up to 120 km. Twenty-four hour broadband communications availability is provided and the network is illustrated in Figure 8.9. The aircraft intercommunicate between HALO gateways, high capacity business premises, and domestic consumer premises. There is a network operations center and also connections out to remote metropolitan centers.

For individual subscribers this type of system should readily provide 1.5-Mbps Internet connections at a price tag of around $40 per month early in the twenty-first century. In the case of dedicated commercial customers, connect speeds of up to 54 Mbps are promised and total network capacity exceeds 16 Gbps.

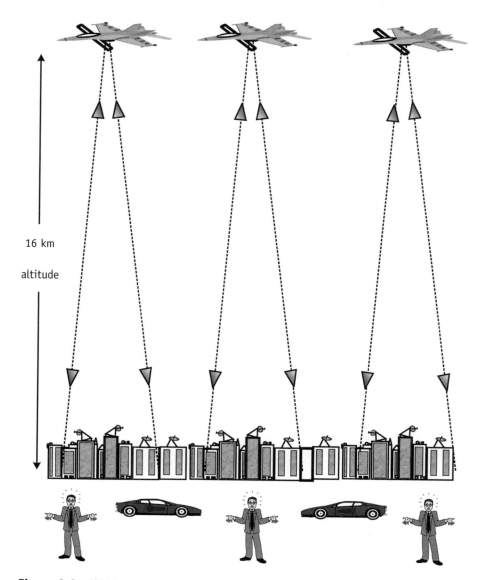

Figure 8.8 HALO-Proteus aircraft in flight above a city area.

Los Angeles will be the first city to have an Angel Technologies' network and 30 further metropolitan areas are targeted in the United States. Other regions of the world will also almost certainly become recipients of

Figure 8.9   HALO network. (Courtesy of Angel Technologies Corp.)

these wireless networks. Compared with terrestrial networks, HALO offers the following advantages:

- Rapid deployment;
- Ubiquitous coverage from the first day of operation;
- Avoidance of local zoning restrictions relating to tower construction;
- Line-of-sight communications to almost all rooftops in each city;
- A flexible and readily upgradeable network.

Also, compared with satellite systems, HALO has the following advantages:

- Between 20 and 2,000 times closer proximity to the users;
- An order of magnitude greater available electrical power on-board;
- Substantial capacity can be allocated directly over densely populated regions;

- Again, a flexible and readily upgradeable network;
- Financing is on a one-market-at-a-time basis.

A direct result of the relatively close proximity referred to above, the round-trip time delay is similar to long-range terrestrial networks, being of the order of tens of microseconds (unlike the many milliseconds for LEOs). This is a highly significant advantage.

With high-flying long-duration manned aircraft, the main challenges are most probably not technological, but human. It will take considerable skills and resourceful approaches to somehow ensure that aircrew can be attracted, appointed, and retained without becoming insane as a result of boredom during completely uninteresting and repetitive 16-hour flights. A major key leading to the solution of this problem will be to provide plenty of absorbing things to occupy the crew at the same time as ensuring that their attention can be rapidly redirected when maneuvers demand this and in the event of any emergency.

Again, future progress with unmanned, remotely piloted, automatically navigated aircraft could eventually enable such vehicles to act as the HALO platforms.

8.6 Technological developments

Composite materials, including carbon fiber, are used in nearly all of the stratospheric vehicles, balloons, and aircraft because of their relatively high strength-to-density ratios.

On the electronics side the same types of advances that have been described elsewhere in this book mainly apply here also. The availability associated with millimeter-wave components and subsystems consistently expands, and costs continually fall. With volume deployment of many of the systems described above unit prices will decrease further. Several of the types of antennas involved are visible in Figure 8.7 and Figure 8.8, and new developments in flat-panel and planar arrays are seen on a regular basis.

For LMDS and MVDS applications many corporations offer highly integrated solid-state subsystems. One example is the transceiver module shown in Figure 8.10.

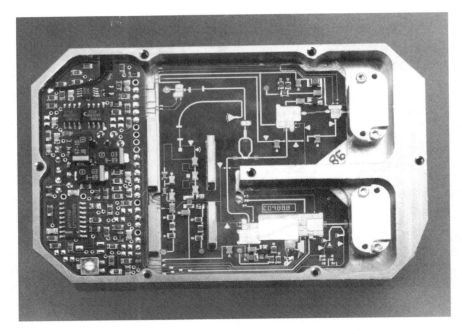

Figure 8.10 A fully assembled 24–26-GHz point-to-multipoint transceiver module. (Courtesy of P-COM Inc. U.K. and Thomson-CSF Microélectronique.)

This 24–26-GHz point-to-multipoint transceiver was jointly developed by P-COM (U.S.-headquartered, but this development is from the R&D unit in the United Kingdom), Thomson-CSF Microélectronique (TCM France), and United Monolithic Semiconductors (UMS France). This module permits full two-way communications and is being manufactured in high volume. The Ka-band portions are completely realized using a total of eight GaAs MMICs and the receiver noise figure is 3 dB over the entire 2-GHz bandwidth.

Hewlett-Packard has developed a 2W solid-state power MMIC amplifier chip for LMDS hub transmitter applications covering the 27.5–28.4-GHz band, that is, a 900-MHz bandwidth. This is a high-linearity chip that can also be adapted to satellite power amplifier applications in the Ka-band. Also, TRW offers a 2.4W high-efficiency MMIC that operates over the 27.5–29.5-GHz range, amounting to a 2-GHz bandwidth.

In production terms, this essentially defines the state-of-the-art regarding critical solid-state high-linearity power amplifier capabilities: up to 2.4W output power over the 27.5–29.5-GHz Ka-band, or even reportedly as high as 36 GHz.

With several terrestrial and stratospheric systems this can often be sufficient power for effective communications to be established, even in adverse weather conditions. Where more power is required, discrete semiconductors can be used as an extra final amplifying stage and up to approximately 4W is achievable this way. Still more power is available from electron tubes such as the traveling wave tube (TWT), and many terrestrial transmitters use such tubes.

The HALO network described in Section 8.5 above does not require any really new technologies. Instead HALO uses equipment readily available from other types of systems, notably LMDS (28 GHz) and also mobile terrestrial wireless infrastructure links at 38 GHz.

8.7 Summary of suborbital platform approaches

In this chapter we have examined a variety of terrestrial and stratospheric technologies and topologies that are being implemented for broadband two-way multimedia applications.

It should be clear that, despite the proliferation of fiber optics and satellite systems, the future also looks bright and expansive for these terrestrial and stratospheric approaches.

Among the terrestrial options introduced in Section 8.2, LMDS and MVDS are the most significant in this context.

As noted above, the specific benefits of LMDS for broadband access are the relatively:

- Rapid deployment;
- Fast ROI and service revenue onset;
- Low network maintenance, management, and operating costs.

With LMDS the operating costs tend to shift away from the infrastructure and toward the CPE.

Another important advantage with both LMDS and MVDS is that these systems are designed on a "grow-as-you-go" basis, so that that none of the system installer's capital becomes wasted when customers churn.

Platforms using stratospheric balloons and high-flying aircraft are also strong contenders for a substantial presence in the broadband communications markets. Among these, Sky Station (balloons) and HALO (high-flying HALO-Proteus aircraft) lead the way, but highly automated unmanned high-flying aircraft platforms may well be the future.

Important technological developments include higher-performance and more compact antennas, power MMICs (up to at least 5W at 36 GHz), and highly integrated transceivers for Ka-band.

It would appear virtually certain that neither single-mode fiber-optic cabled networks nor broadband satellite systems will have the future all their own way—not by any means.

Select bibliography

Angel Technologies Corporation, Press Releases 1999; http://www.angelcorp.com and http://www.broadband.com.

"Cisco and Motorola in telecoms joint buy," Editorial item in *Electronics Times* (U.K.), June 16, 1999.

"Commute-I raises hopes for MVDS," Cable Europe, November 25, 1998, p. 5.

Heftman, G., "LMDS Set To Challenge For Last-Mile Supremacy," *Microwaves & RF*, April 1999, pp. 30–38.

Drake-Wilkes, J., (GEC-Marconi), "Multimedia Distribution Technology," (presentation material personally supplied to the author).

Shu, J., et al., "Ka-band 2 Watt Power SSPA for LMDS Application," *IEEE MTT-S Digest*, 1998, p. 573.

Siddiqui, M., et al., "A High Power and High Efficiency Monolithic Power Amplifier for Local Multipoint Distribution Service," *IEEE MTT-S Digest*, 1998, pp. 569–572.

Steckley, E., "Broadband Wireless—The Dawn of a New Era," *Telecommunications*, February 1998, pp. 51–57.

"US in microwave comms go ahead," Editorial item in *Electronics Weekly* (U.K.), February 25, 1998.

Viaud, J.P., et al., "Modular Millimetre-wave transceiver design for today's digital radio links," *Microwave Engineering Europe*, February 1999, pp. 49–54.

Wood, L. "Lloyd's satellite constellations," WWW, e-mail: L.Wood@surrey.ac.uk.

9

Broadband Mobile Satellite Systems

9.1 Third-millennium satellites

Through the middle years of the twentieth century most national and international communications were truly wired. Armored copper cables containing up to hundreds of copper wires covered the industrialized countries and crossed oceans beneath the sea. These cables enabled several simultaneous telephone conversations to take place via electro-mechanically switched exchanges (telephone switching centers), and until well after World War II these cables served communities satisfactorily.

The exception to cables was short-wave high-frequency (HF) radio, but this was unreliable and only enthusiasts (and the military) made any serious use of the medium.

Today satellites are a part of everyday life, with TV delivered this way probably being the most visible aspect for the majority of people. The basic concept of satellite communications has been with us for well over half a century. Things began with Arthur C. Clarke's groundbreaking (space-breaking?) "Wireless World" paper published in 1945 in which he showed that three satellites placed in geostationary orbit would have footprints covering about 90% of the earth's surface. *Geostationary* means that the spacecraft's orbits and orbital speeds make them appear stationary with respect to the corresponding point on the earth's surface immediately below the orbiting spacecraft.

Following World War II, European reconstruction and other more fundamentally pressing needs demanded international attention. As a result it was many years before technology reached the point where it became feasible to seriously consider satellite communications as a practical possibility.

In 1957 the Soviets launched the first artificial satellite (Sputnik) and the space race was well and truly on. On June 10, 1962, the first Telstar communications satellite was launched—paving the way for space-based international communications on a wider scale. This first-ever Telstar was placed into an unusual low earth orbit (LEO). The spacecraft for this 1962 project had an orbit varying from 800-km to 4,800-km altitude, and only the relatively near earth segment, when the satellite was around 800–1,000 km up, was usable for communications. The Telstar satellites were eventually purchased by AT&T and this explains why the name has been continued up to the present day.

This first Telstar operated on a bearer frequency of 4.17 GHz and the transponder could either relay one (yes—just one) TV channel or alternatively 60 simultaneous phone conversations. This may all seem tame by third-millennium standards but the technology was truly revolutionary in 1962. Remember, this was before integrated circuits existed at all; it was when any low-powered computer would fill a large office suite, and even transistors were "pushed" to work at frequencies into VHF. The spacecraft for this program was designed and assembled by Bell Labs (now owned by Lucent Technologies). But the first Telstar proved that one could actually relay radio communications through an orbiting spacecraft and it was followed by various other experiments such as Project Relay and Early Bird.

Arthur C. Clarke's concept of the geostationary earth orbit (GEO) and wide earth coverage remained, however, the linchpin that would lead to most of the progress in the 1970s and 1980s. Indeed, only a year after the first Telstar, the first GEOs were launched by NASA and Hughes and were known by the generic name *Syncom*. Nowadays international trunk satellite communications, served by Intelsat and others, and also services like DTH and DBS (satellite TV) all use GEOs.

The types of satellite orbits suitable for communications applications are shown in Figure 9.1. It should be noted that three orders of magnitude are involved here, with Iridium at 780 km altitude, Globalstar at 1,400 km, and all the others occupying orbits in the 10,000–36,000-km range.

The LEOs, typified by Iridium and Globalstar here, appear to race around the globe because of their proximity, while at the other extreme GEOs (geosynchronous or geostationary spacecraft) appear as though they are fixed in the sky. Global positioning satellites (GPS, GNSS, GLONASS) are, circling at 20,200 km, considerably lower than GEOs because they must retain some relative motion in order to provide the navigational information from a multiple-satellite configuration.

Transmission delays with all the LEO, HEO, and MEO (highly elliptical and medium earth orbiting) systems are much lower than for the GEOs, being mostly in the tens of milliseconds range. In contrast GEO users suffer a round-trip transmission time delay of 540 ms, and this is very noticeable when using a GEO-based speech channel. Echo cancellers are regularly installed to eliminate echo in these types of links. On the other hand, GEO-based systems involve negligible Doppler shift problems because of their geostationary character.

Although LEO orbits, in particular, have relatively small transmission time delays, the problem of Doppler shift is much greater than with GEOs because of the substantial differences between satellite motion and terrestrial receivers. It has only been practical to correct for this Doppler shift with the advent of advanced electronic systems. Link power budgets are greatly eased with LEO systems because of the much closer proximity to earth, allowing all solid state technology to be implemented increasingly even at the higher frequency bands.

In this chapter we are to some extent concerned with all types of satellite constellations.

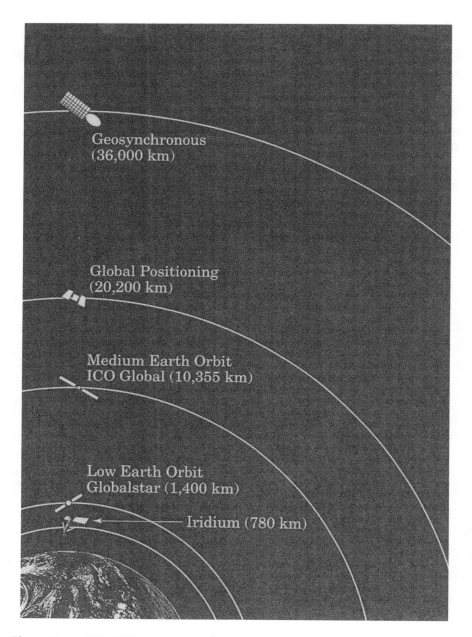

Figure 9.1   GEO, MEO and LEO satellite orbits. (*From*: Nellist, J.G., and E.M. Gilbert, *Understanding Modern Telecommunications and the Information Superhighway*, Norwood, MA: Artech House, Inc., 1999.)

It is well known that around 200 GEO satellites are currently in orbit, with even more planned, and also that several LEO and HEO satellites are either in orbit or at the planning stage.

A note of caution is in order here. Organizations that are in the vanguard (i.e., first in) with new technologies frequently suffer more than later entrants in terms of early financing and return on their investment (ROI) over the short term, that is, the early years. By mid-1999 it seemed that this was applying to the Iridium LEO project because, although Motorola had this fully up and running, the subscriber take-up was much slower than anticipated.

At the time of this writing, Iridium's major competitor, Globalstar, is not yet operational and it will be interesting to observe the effects that the two competing operators have on the market. Both of these projects only really provide narrow-band services as a result of the limited spectra available (around 4 MHz), and in some ways they are "proving programs." Available power is also relatively low. When LEOs were first planned and designed the explosive growth of the Internet was not envisioned and voice services remained dominant.

With new technologies, especially on this large scale, operators often have to "fly by the seat of their pants." However, this scenario has certainly not inhibited potential providers of broadband satellite services that are expected to provide much faster Internet access via both GEO and LEO systems.

9.2 Some broadband mobile satellite systems

The near future is almost certainly with broadband and increasingly mobile satellite services, competing with and complementary to terrestrial and stratospheric services (see Chapter 8), and also of course DWDM fiber systems (see Chapter 3).

Several forecasts have been made for broadband satellite IP service markets and one is shown in Figure 9.2. This data was originally generated by Merrill Lynch & Co. and also reproduced in *Communications Week International* in May 1999. Through the turn of the millennium annual growth in this sector is over 100%, which is nothing short of phenomenal. Major recipients of the rewards from this growth include LoralOrion and

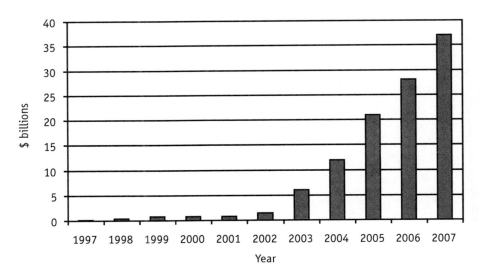

Figure 9.2   Global markets for broadband (IP) satellite services.

PanAmSat, and the global market is expected to be worth around $37 bil-
lion in 2007.

In regions of low subscriber density, satellite systems will become the
dominant means of providing both voice and Internet services. With relatively dense and developed centers, the various delivery mechanisms will
compete heavily—with costs, maintenance, and time-to-deployment
being critical factors. It is important to note that mature satellite systems
can usually be deployed much more rapidly than cables can be laid and
brought online.

The best known names in the broadband satellite scenario are Hughes
(Galaxy/Spaceway), Lockheed Martin (Astrolink), SkyBridge, and, of
course, Teledesic. Because of their relatively high profiles and prospects,
we provide substantial information on these four projects here.

However, they are far from the only programs that have been planned
and some details on a selection of broadband systems are given in
Table 9.1.

Most proposed systems use the Ka-band but some operate using the
lower Ku-band. SkyBridge is an interesting example of the latter that is
discussed presently when this player is considered in more detail.

Table 9.1

Features Applying to a Range of Proposed Broadband Mobile Satellite Systems (MSS) (Kruger, West, Williamson, Wood, and Others)

System Name	Number of Satellites and Configuration	User Data Rate	Investment $M
Alenia Spazio EuroSkyWay	2 in first phase 5 in second phase (GEO)	144–2,048 Kbps (up) 32.768 Mbps (down)	500 first phase
GE Americom Ge Star	9 (GEO)	384 Kbps–40 Mbps	4,000
Hughes Galaxy/Spaceway	9 (GEO)	384 Kbps–6 Mbps	3,200
Lockheed Martin AstroLink	9 (GEO)	384 Kbps–8.448 Mbps	2,400
LoralOrion (LAHI) CyberStar	3 (LEO)	384 Kbps–3.088 Mbps	1,050
Motorola Millennium	4 (GEO)	384 Kbps–51.84 Mbps	2,340
SkyBridge	80 (LEO)	0 Mbps (commercial) and 2 Mbps (residential)	4,100
Teledesic	288 (LEO)	64 Mbps	9,000

From the all-important user viewpoint it is the downlink bit rates that are so significant. Most of the examples summarized in Table 9.1 show systems with downlink capabilities exceeding 30 Mbps. Where this is not the case, as with Hughes Galaxy, Astrolink, and LoralOrion's Cyberstar, in most instances further project plans have been filed with the FCC with a view to providing higher bandwidth constellations. This is being driven by increased competition and user expectations in coming years. The four main broadband MSS and one regional MSS are now considered in more detail.

9.2.1 Hughes Galaxy/Spaceway

The correct name for the Galaxy project is really the $3.2 billion Space-way project, from Hughes, and this comprises a GEO constellation which has all but one satellite linked with intersatellite links (ISLs). A nine-satellite constellation is proposed with the capability of up to 6-Mbps access on the downlink. An artist's impression of a typical "Galaxy" spacecraft is shown in Figure 9.3, in which the phased array antennas are clearly visible as square, hexagonal, and octagonal elements surrounding the payload in the center.

Hughes has used the name *Galaxy* for several projects over the years, including most of their satellites; however, it is important to refer to the broadband program now known as "Spaceway" in this context. Indeed, it is planned for the Galaxy name to be retired as a brand.

In 1999 Hughes filed with the FCC for an entirely new concept called Expressway. This would deploy 14 GEO satellites in 10 locations, linked with optical ISLs, and having an estimated cost of $4 billion. Expressway would use a combination of Ku-band and the millimetric V-band rather than Ka-band, providing much greater capacity than most of Hughes' rivals.

As discussed in Chapter 2, the 40–50-GHz V-band involves modest atmospheric attenuation but the effects of precipitation are considerably worse than with lower frequencies such as Ku-band or even Ka-band.

Interestingly, early in 1999 Hughes also filed for a MEO constellation comprising 20 active satellites.

9.2.2 Lockheed Martin (Astrolink)

Lockheed Martin Corp., Telecom Italia SpA, and TRW Inc. have announced a plan to invest $900 million in Astrolink LLC, a venture that plans to use satellites for high-speed communications and Internet access.

Lockheed Martin, the second-largest U.S. aerospace company, will own 46% of Astrolink, investing $400 million, with the remainder split between Telecom Italia, Italy's biggest phone company, and TRW, a U.S. aerospace and vehicle components maker. Each company will invest $250 million in this project.

Astrolink, like competitors Teledesic LLC and Hughes Electronics Corporation's Spaceway project, plans to use satellites for linking computer networks worldwide. Astrolink plans to launch its first satellite in 2002 and to begin commercial service in 2003. The project's total cost is estimated at $3.6 billion.

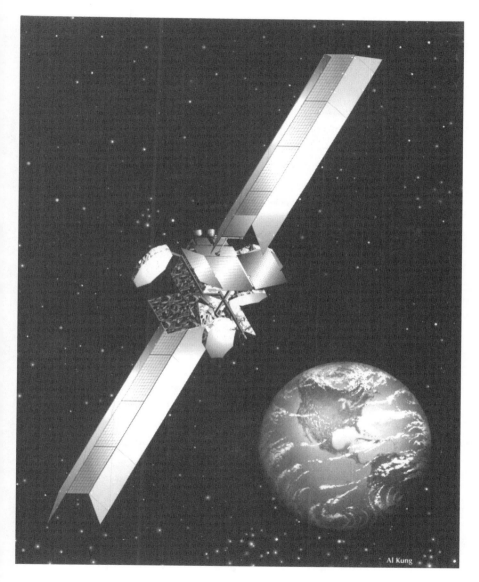

Figure 9.3 An artist's impression of a "Spaceway" spacecraft. (Courtesy of Hughes Electronics Corp.)

Lockheed Martin will provide the satellites and the launchers; Telecom Italia and its Telespazio unit will contribute ground control systems and the marketing of services in Italy, Europe, and Latin America; while TRW will provide the communications transponders.

9.2.3 SkyBridge

Competing strongly with Teledesic, SkyBridge is an unusual project in that it uses the Ku-band (lower than most other operators' Ka-band) and employs frequency reuse adaptively in order to offer real broadband services. Through the adoption of Ku-band and frequency reuse, the utilization of the radio-frequency spectrum is also optimized as well as protecting GEO and terrestrial communications systems using the same frequencies.

SkyBridge, which is an Alcatel Company, went for Ku-band because of the availability of proven and less expensive technology at these lower frequencies. Other features include:

- The absence of any intersatellite links (no ISLs);
- All switching ground-based;
- Use of the standard ATM protocol (see Chapters 1 and 2).

The absence of ISLs combined with all the switching occurring on the ground should increase the overall reliability and robustness of the system at the same time as keeping costs down below $4.2 billion. However, it limits connectivity and requires that all communications pass through hub earth stations.

This system is being designed to offer up to 20 Mbps to commercial customers and up to 2 Mbps to residential customers, representing a high-speed Internet on-ramp capability. Any multiple of this capacity will be made available to business users on demand and, overall, a total global market of over 20 million users is confidently expected.

By 1998 a large industrial team was in place under the leadership of Alcatel to design and develop SkyBridge. More than 400 engineers were working on the program and these large-scale engineering activities enabled SkyBridge to finalize the design characteristics of their system.

An artist's impression of a typical SkyBridge spacecraft is shown in Figure 9.4 and part of the constellation is shown in Figure 9.5. In Figure 9.4 the antennas comprise specially designed radiating arrays fed

Figure 9.4 An artist's impression of a SkyBridge spacecraft. (Courtesy of Barrington Lloyd Intl., U.K.)

Figure 9.5   An artist's impression of the SkyBridge constellation. (Courtesy of Barrington Lloyd Intl., U.K.)

by beam forming networks and these are briefly described later in this chapter.

The overall systems architecture is indicated in Figure 9.6 where some satellites of the constellation are shown on the left-hand side while the architecture itself is shown to the right. Broadband signals are beamed up and down between space and ground using the high data rate infrastructure links. The main ground station antennas connect to ATM switches to form the local gateways. ATM is discussed in Chapters 1 and 2.

Practically any type of terrestrial networks can then be attached to these ATM switches: servers, narrowband networks, and broadband networks that are further attached either directly to users or to other servers.

Alcatel is the general partner of SkyBridge L.P., incorporated as a limited partnership in Delaware, U.S. The other partners of Sky-Bridge include the following group of leading industrial companies: LoralOrion of the United States; Toshiba Corporation, Mitsubishi

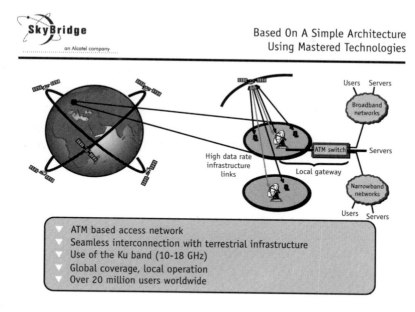

Figure 9.6   Overall architecture of the Skybridge system. (Courtesy of Barrington Lloyd Intl., U.K.)

Electric Corporation, and Sharp Corporation of Japan; SPAR Aerospace Limited of Canada; Aerospatiale and CNES of France; and SRIW, a Belgian investment entity.

At the time of this writing a challenging dispute is under way between SkyBridge, PanAmSat, and others concerning frequency interference problems. However, according to the latest information from the company, full deployment of the SkyBridge system is expected in 2002.

9.2.4 Teledesic

When the Teledesic system was first proposed it comprised a massive 840-satellite constellation, all with intelligent ISLs. Since that time (mid-1990s) the constellation concept has been reduced by almost three-fold, and as presently conceived, the Teledesic Network will consist of 288 operational satellites, divided into 12 planes, each with 24 satellites.

The satellites orbit in planes having north-to-south and south-to-north orientations, with the earth rotating beneath this network.

Teledesic is probably the best known of all the broadband projects and there are two reasons for this. The first is probably just by being "first in," as described above. The second and very important reason is doubtless inextricably linked to the fact that none other than Microsoft's Bill Gates joined with Craig McCaw to initiate this venture. The contract team was originally composed of Boeing and Matra but later expanded to include Motorola. However, this was under review at the time of this writing.

Figure 9.7 is an artist's impression of a Teledesic spacecraft, dominated by the large solar arrays for electrical power. The phased array antennas are facing earth and therefore hidden from view in this figure. The full constellation of 288 satellites is depicted in Figure 9.8 and an impression of the ground (and air) segments is illustrated in Figure 9.9. Interactive signals to and from Teledesic may be relayed between the network, mobile, or stationary ground and sea nodes; aircraft; and other

Figure 9.7   An artist's impression of a Teledesic spacecraft. (Courtesy of Teledesic LLC.)

Figure 9.8 An artist's impression of the Teledesic constellation. (Courtesy of Teledesic LLC.)

spacecraft. It is also possible in the future, given suitable gateways and handover protocols, that communications between other competing space-based networks may be available.

The average latency (the time delay as the signal travels through the uplink and then the downlink) is 70 ms with Teledesic, and the organization is anticipating that around 750,000 small-to-medium-sized businesses in the United States alone will go for these broadband services. At the end of the twentieth century only 3% of such businesses had access to any level of fiber connection and the expectation is that this situation will only slowly change over the early years of the third millennium. Fiber connections are discussed in Chapters 3 and 6.

As planned, the Teledesic system will use Ka-band with a bearer frequency of 30 GHz for the uplinks and 20 GHz for the downlinks.

Originally it was anticipated that Teledesic would become operational by the year 2000, but this became less likely when the management announced a funding program covering a period of four years. By the year 2002 or 2003 this advanced "space-based Internet II" facility is expected to be functioning commercially.

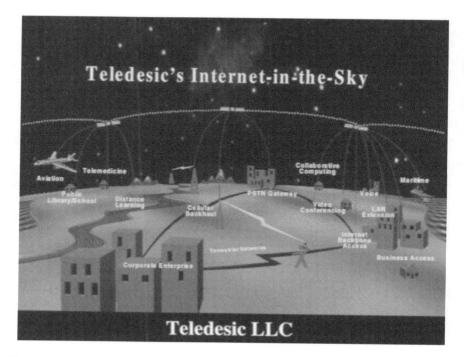

Figure 9.9   Overall architecture of the Teledesic system. (Courtesy of Teledesic LLC.)

Using a standard Teledesic terminal, which supports a 64-Mbps downlink, Figures 9.7–9.9 (i.e., including the photographic figures—but in full color) could be downloaded in less than a second.

9.2.5 Thuraya

Although not strictly a broadband MSS this is an important turnkey project that is designed to serve a wide region centered upon Abu Dhabi in the United Arab Emirates. Within this country the operation will be operated and run by Thuraya Satellite Communications Co. Ltd., a company headquartered in Abu Dhabi.

The population approaches 2 billion people living in the region that will be served by the Thuraya mobile communications system that is scheduled to begin operations in the year 2000. Hughes Space and Communications International, Inc., (HSCI) and Thuraya Satellite

Telecommunications Company of United Arab Emirates signed a contract on September 11, 1997, for the $960 million turnkey system. This price includes the manufacture of two high-power HS-GEM (GEO-mobile) satellites with the launch of the first spacecraft planned to take place early in the year 2000. Other features included are insurance, ground facilities, and the provision of user handsets. The second spacecraft will be a ground spare, and there is an option for a third satellite. Thuraya's footprint encompasses the Middle East, North and Central Africa, Europe, Central Asia, and the Indian subcontinent.

HSCI's team includes Hughes Network Systems, Inc. (HNS) and the spacecraft are being built at Hughes Space and Communications Company's Integrated Satellite Factory near Los Angeles International Airport. HNS is also providing the ground facilities, while both HNS and Ascom of Switzerland are providing a total of 235,000 handsets.

The Thuraya satellites are the first in Hughes' HS-GEM series. This product line expands Hughes' offerings beyond satellite manufacturing, making the corporation a large-scale satellite systems integrator. The spacecraft constellation integrates a high-power GEO satellite (derived from the HS 702 body-stabilized design) with ground segment and user handsets, to provide a range of cellular-like services over a large geographic region. The Thuraya ground segment includes terrestrial gateways plus a collocated network operations center and satellite control facility in the United Arab Emirates.

Thuraya offers GSM-compatible mobile telephone services, transmitting and receiving calls through a single 12.25-meter-aperture reflector. The satellites employ state-of-the-art on-board digital signal processing to generate more than 200 spot beams that can be automatically redirected to predetermined earth locations, allowing the Thuraya system to adapt to business demands in real time. Calls will be routed directly from one handheld unit to another, or to any terrestrial network, and the system has the capacity for 13,750 simultaneous voice circuits.

An artist's impression of a typical Thuraya spacecraft is shown in Figure 9.10 where the large mesh reflector unit is a dominant feature. The phased arrays comprise 128 active elements with 17W solid state power amplifiers (SSPAs) and the communications payload module also has 125W traveling wave tubes (TWTs) on board.

Figure 9.10   An artist's impression of a typical Thuraya spacecraft. (Courtesy of Hughes Electronics Corp.)

9.2.6 Some further proposed broadband satellite systems

This is not intended as anything like a complete listing of proposed broadband MSS—it never could be because industry realignments take place at frequent intervals and various programs become subsumed into other projects. The following information should therefore be treated only as indications regarding further programs that have been announced.

EuroSkyWay is planned for full operation in the year 2002 as Europe's first broadband MSS exclusively reserved for multimedia services. It is a five-GEO satellite constellation from Alenia Aerospazio which is owned by Finmeccanica of Italy. Initially only two satellites will be launched for basic testing purposes, but once the second and final five-satellite phase is completed users will have access to a 32.768-Mbps downlink facility. The specialty car maker Ferrari is equipping a demonstration vehicle (a Maranello road car) with multimedia and navigation technology that will enable non-driving occupants to download and handle all forms of traffic information, book accommodation, and watch films.

GE *StarPlus is a nine-satellite GEO system planned by GE Americom that would be capable of providing 40 Mbps of user bit rate access.

Lockheed-Martin's Astrolink is also a nine-satellite GEO system that would operate in V-band and would implement ISLs. Users of this system would be able to receive and transmit data at up to 8.448 Mbps.

LoralOrion have announced their Cyberstar program and also their much more advanced and probably therefore more speculative Cyber-Path (V-band) systems. Cyberstar is a three-satellite LEO that limits the users to just 3.088 Mbps, but CyberPath would increase this by an order of magnitude.

Motorola's "Millennium" project comprises a four-LEO constellation operating in V-band and enabling users to access the Internet at speeds up to 51.84 Mbps.

Then there is the Ellipso project, with the spacecraft being manufactured by Orbital Sciences, Inc. This is mentioned here because Orbital Sciences have announced that they propose an extension to the basic Ellipso configuration, named Orblink, conceived to provide a broadband (V-band) facility.

Ellipso does not provide global coverage but instead concentrates on regions that are likely to provide a market sufficient to enable a profitable ROI to be achieved. Boeing is a major investor in Ellipso and achieved the substantial hurdle of getting FCC approval in July 1997.

PanAmSat has a project plan named V-Stream that comprises a 12-satellite GEO broadband MSS using V-band bearers.

Pentriad is a proposed primarily V-band constellation using thirteen satellites in highly elliptical (HEO) Molniya orbits over the Northern Hemisphere (Molniya orbits refer to those used traditionally by the former United Soviet Socialist Republic, now the Commonwealth of Independent States). Pentriad has also stated that they desire to provide a wireless infrastructure for the HALO broadband stratospheric air-to-ground communications project described in Chapter 8.

Another planned project is GESN, announced by TRW. This is a hybrid MEO/GEO project the working name of which is *Global EHF Satellite Network* (or GESN). It comprises fifteen MEO satellites at the same altitude as ICO[1] and the now-defunct Odyssey, and four GEO satellites. Latest indications from TRW are that this would be a Ka-band system.

Two other proposals are worth mentioning here.

One is termed *GIPSE*—an in-house system at Surrey University in the United Kingdom. The aim of GIPSE is to design a satellite constellation capable of providing broadband multimedia personal communications.

There is also a Russian constellation proposal and this again amounts to a hybrid system, being a combination of LEO and MEO constellations. There would be seven planes of thirteen satellites apiece orbiting at 700-km altitude (the N-system) and four additional planes of six satellites apiece orbiting in a MEO configuration (the V-system).

9.3 Prospects for these space-based Internet on-ramps

It is extremely unlikely that all or even most of the broadband satellite systems described in Section 9.2 will ever get built.

We have seen the subsuming processes relating to earlier programs such as Celestri and M-Star and it would appear likely that many of the remaining planned projects will literally never get off the ground. However, competitive positioning is constantly proceeding and when the necessary engineering planning work is done some of this will cross-fertilize to updated, more realistic, and timely programs.

---

1. In late August 1999, ICO filed for bankruptcy.

Figure 8.1 of Chapter 8 shows the forecast for the numbers of subscribers, globally, to all forms of broadband services. It is clear that the third millennium (i.e., beginning with the twenty-first century) will most probably be seeing subscriber numbers amounting to hundreds of millions and growing strongly. Some of these, perhaps around a quarter of them in 2001, will go for broadband satellite. A major factor in this choice will be the need to have a fast Internet connection while not being in the frame for either LMDS or fiber, described in Chapters 8 and 3, respectively.

As recently as 1999 at least one estimate put the proportion of U.S. Internet users with a fiber access facility as only 3%. While this percentage will doubtless increase eventually, in the meantime broadband wireless connections are the only choice in such instances—and then there is the mobile dimension.

Given the current intense commercial interest combined with rapid and comparatively low-cost deployment, LMDS and similar terrestrial systems will compete heavily in many regions at an early stage.

By assuming that approximately one-third of the totals are broadband satellite in 2001 but having the proportion grow considerably later, it is possible to roughly estimate subscriber numbers and growth over the 1997–2007 time frame. The results of this process are shown in Figure 9.11 where it is clear that competition from both terrestrial and

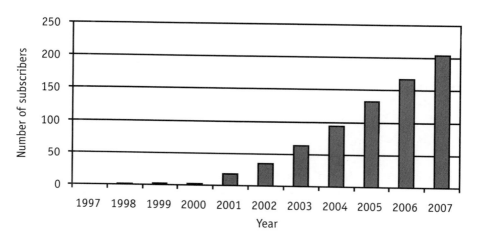

Figure 9.11   Global subscribers to broadband satellite services.

stratospheric installations tends to restrict the proportion for MSS to around 34% of the total in 2005 although growing steadily over the 2005-plus time frame.

However, an important advantage of MSS is the inherent capability for mobile access—as illustrated by the EuroSkyWay/Ferrari demonstration referred to above. Only the stratospheric Sky Station and HALO types of configurations really compete with this capability—and HALO is metropolitan region–oriented whereas MSS is "go anywhere."

9.4 Space segment technologies

By the end of the twentieth century considerable masses of spacecraft were regularly lifted into space by commercial unmanned rocket launchers or occasionally the space shuttle. The efficiencies of the solar cell arrays for converting sunlight into electricity were steadily improved—and more array area was deployable in space. Therefore, substantial amounts of electrical energy were available and (noting, for example, the Thuraya spacecraft) active amplifying components like 125W TWTs could function commercially in orbit.

Where tubes have to be used they are driven by solid state power amplifier modules (SSPAs). The microwave or millimeter-wave (mm-wave) signal is inputted to the SSPA that provides a power output level ranging from somewhat less than one watt up to several watts. This is sufficient to linearly drive a high power amplifier such as a TWT, the output of which feeds the antenna. Both the SSPA and the TWT are supplied with electrical power from an electronic power conditioner (EPC). The overall arrangement is shown in Figure 9.12.

In a spacecraft the main power supply comes from the solar cell array and the EPC condition is continuously controlled and monitored.

Also by the end of the twentieth century large antenna arrays, including substantial phased arrays, were already practical propositions for satellites, and significant payloads were feasible embodying much digital processing and microwave or millimeter-wave electronics. Phased arrays with active elements (SSPAs) only requiring some tens of watts per element are now being regularly implemented.

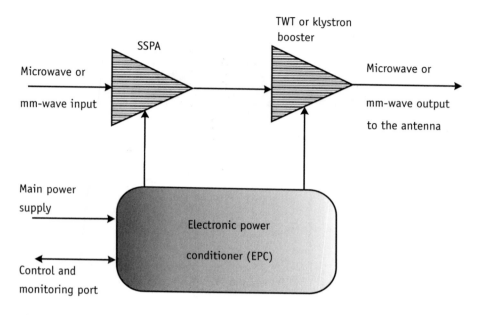

Figure 9.12 Microwave/millimeter-wave power module (MPM).

However, space remains essentially a hostile environment, and all these components must be designed, manufactured, and tested "fit for space." Also, in spite of the advances in launch capabilities and orbital injection techniques, the all-up weight and physical dimensions of the entire spacecraft remain a significant consideration.

These comments apply equally to all types of electronic components and subsystems aboard spacecraft.

Apart from phased arrays several novel types of antennas and signal feeds have been designed. One example is that of the SkyBridge spacecraft antenna and the overall structure of this is indicated in Figure 9.13. This antenna is based in principle upon a design known as STENTOR already deployed in Ku-band GEO spacecraft.

The "cutting edge" of the SkyBridge antenna is the radiating array with its 126 elements (for simplicity only 82 are shown in the schematic of Figure 9.13). This array would just fit within a 25 cm × 25 cm × 2 cm box. From each antenna element eight spot beams are required to be dynamically configured using beam forming techniques and filters.

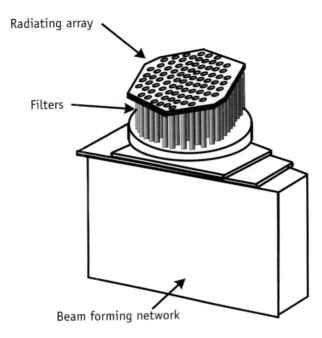

Figure 9.13   SkyBridge spacecraft antenna.

The entire signal to be downlinked is fed, within the spacecraft transponder, into the beam forming network (BFN). The 126 outputs from this are then fed into filters behind each radiating element. These filters separate the signal paths to the required elements in the correct phases in FDM mode (see Chapter 2 for FDM).

Because the BFN inherently involves separation of the input signal into the 126 outputs, several amplifiers are required within this portion of the system. Losses within the BFN, mainly due to the power separation, are specified to be less than 57 dB, and each SSPA amplifier must have a power gain of over 60 dB to boost the signal. These SSPAs are also required to have at least 500-MHz bandwidths.

Each SkyBridge spacecraft has 100 BFNs each with no less than 1,134 MMICs and over 10,000 SSPAs. The technology chosen for these SSPAs is multichip modules (MCMs) in which many MMICs and other ICs are hybrid integrated within single packages. This is quite an extreme example and most other systems will have far fewer MMICs and SSPAs.

9.5 Ground segment technologies

Back on the ground, the requirements, the environment, and the possibilities are, of course, entirely different from those applying to space.

The requirements within the ground segment are generally much more complex. There are fixed and mobile users, wide-scale distribution, and all the systems demands of an architecture similar to that depicted in Figure 9.6 for SkyBridge with 80 spacecraft—or with Teledesic that will have 288 spacecraft interacting with a network of at least this complexity. There is also precipitation attenuation to contend with on all wireless links and not just the earth ends of the satellite-to-ground connections.

Otherwise, however, things are relatively benign. On earth there is usually plenty of room and plenty of energy for supplying every element of the system. Maintenance is also quite straightforward since in most places technicians can be dispatched to deal with problems as and when they arise.

Earth stations for broadband satellite systems have similar outward appearances to most other facilities, including those for narrowband installations. An example of a large state-of-the-art earth station, located in Papua New Guinea, is shown in Figure 9.14. Another example (see Figure 9.15) is the satellite downlink facility for the ABC Broadcasting Company in Australia. These photographs are of Scientific-Atlanta facilities, but corporations such as Hughes Electronics, for example, also provide earth stations.

Uplink power amplifiers in earth stations have similar configurations to that shown in Figure 9.12 and the description provided for that MPM apply equally here, except for the final high-power amplifying tube. On earth, more power can be used, on occasion reaching kilowatt levels and sometimes furnished by a klystron rather than a TWT.

9.6 Concluding remarks

Satellite communications have a long history beginning with Arthur C. Clarke's brilliant GEO concept of 1945, followed much later by the first Telstar and other non-GEO launches, and then the lengthy period of GEO-based trunk communications and, of course, satellite TV (DTH).

Figure 9.14   A satellite earth station located in Papua New Guinea. (Courtesy of Scientific-Atlanta.)

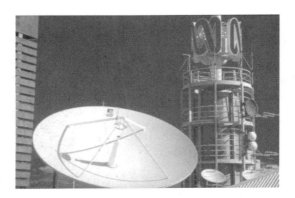

Figure 9.15   A satellite downlink for ABC Australia. (Courtesy of Scientific-Atlanta.)

Following the Iridium and Globalstar narrowband LEO constellations, the broadband satellite era is now imminent with several of the systems described here going live and online—and the prospects for rapid Internet access beckon. As the first decade of the twenty-first century progresses, so all forms of broadband access technologies come into their own globally.

Select bibliography

Bulloch, C., "Mobile Satellite Services Face Delay," *Telecommunications*, October 1998, p. 15.

Elbert, B., *Introduction to Satellite Communications, Second Edition*, Norwood MA: Artech House, 1999.

Foley, T., "Satellite gets Tied up with Cable," *Communications Week International* (CWI), May 10, 1999, pp. 6–7.

Haviland, R., "Broadcasting Satellites," *Discovery, Professional and Industrial Publishing Ltd.*, August 1962, pp. 9–14.

Kruger, P., "Back Down to Earth," *Communications International*, April 1999, pp. 26–27.

Nellist, J.G., and E.M. Gilbert, *Understanding Modern Telecommunications and the Information Superhighway*, Norwood, MA: Artech House, Inc., 1999.

"Satellites for Internet," *Inside Digital TV*, December 14, 1998, p. 5.

"Sky High Target for Active Array Performance," Microwave Engineering Europe, June 1998, pp. 20–23.

"Speed satellite," Editorial item in *Electronics Times* (U.K.), May 5, 1999.

West, Simon, Barrington Lloyd International, information concerning SkyBridge. (N. B. www.skybridge.com is not the Web site for the limited partnership of which Alcatel is general partner, known as SkyBridge L.P. See Alcatel instead.)

Williamson, M., "Can Satellites Unblock the Internet?," *IEE Review*, May 1999, pp. 107–111.

Wood, L., "Lloyd's satellite constellations," WWW, e-mail: L.Wood@surrey.ac.uk.

Web site: www.hughespace.com.

Web site: www.teledesic.com.

10

High-Speed Digital Epilogue

10.1   The multi-trillion dollar business-and-leisure superhighway

In 1998 the global total amount of revenue generated by e-business was estimated at around $500 billion, or half a trillion dollars. This comprised online goods and services and almost one-half of the total arose from the hardware and software consumed in building the infrastructure of the Internet.

Without a shadow of a doubt the Internet is drastically reconfiguring the global marketplace—a trend that will accelerate during the twenty-first century. By the year 2002, the global total amount of revenue generated by e-business alone is expected to approach a trillion dollars, and in subsequent years this will almost certainly grow to several trillion dollars. Services will probably be dominated by business needs. However, leisure pursuits such as active global games will also be played across the Internet (with built-in storage and an understanding concerning time zones).

225

A major factor that will facilitate this expansion is the ever-decreasing costs associated with individual communications channels and this trend is indicated in Figure 10.1. The rate of decrease was clearly highest during the 1940–1980 period and this can be traced to several influences. During the first part of this period point-to-point microwave links were established across most developed countries, and then, from the late 1970s, the powerful combination of fiber-optic transmission and satellite communications forced costs per channel further downwards. By 2020 these costs essentially tend toward zero—the use of communications channels will come free and revenues will be generated by subscription to various services and advertising income.

The infrastructure technologies will add on to the budgets for the future development of the Internet that will extend out to versions II and III. Increasing implementations of broadband connections, including optical and wireless technologies (e.g., 3G mobile), will accelerate this growth.

During the second decade of the third millennium the various traditional approaches, founded during the previous century, will converge and this process is shown in Figure 10.2. The universal Internet, corporate intranets (private communications networks interconnecting many company divisions), CATV operators, and both the telcos and mobile service operators have their functions converging through multimedia

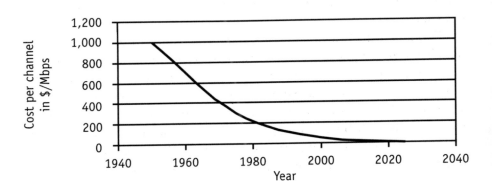

Figure 10.1   Costs per communications channel trend.

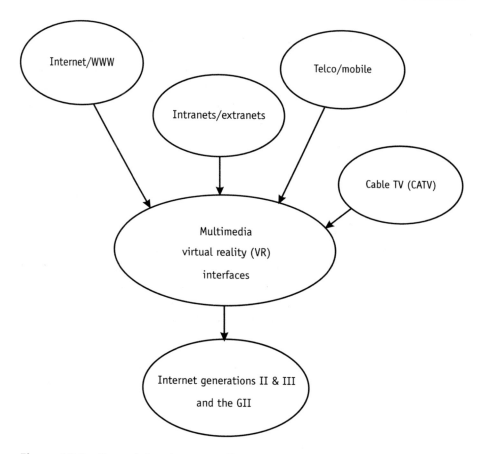

Figure 10.2 Second decade twenty-first–century communications convergence.

virtual reality interfaces. The outcome of this is partly the next-generation Internet (II and then III) and the global information infrastructure (GII). At this point CATV operators will be offering upgraded services based upon the implementation of broadband single-mode fibers.

As discussed elsewhere in this book the actual technologies will be highly varied during the formative years, including many advances in both fixed links and broadband wireless, until optical approaches gradually take over and eventually dominate.

10.2 Gadgets and gizmos 2010 style

By the year 2010 most people will use the Internet with the same casual ease that applied to POTS in the latter part of the twentieth century. It will not be necessary to be at all PC literate in order to communicate using the Web, and people will increasingly possess add-on units, mostly remote-controlled and mobile, for use in conjunction with their digital TV receivers.

Another likely trend is for the mobile phone itself to become an Internet-access device. Obviously this will require a relatively broadband connection directly to each phone in order to enjoy rapid access, but again optical technologies appear to be pointing the way.

There is a precedent for this trend, namely the Minitel that was launched in France during the 1970s and which has enjoyed such wide acceptance and success ever since. The Minitel comprises a phone with a flat-panel screen on which the messages and data are displayed.

Other trends, beginning with the most imminent and ending with the most advanced, could well include:

- Personal electronic goods as fashion accessories;

- Personal electronic and photonic assistants (PEPAs);

- Auto-scan-and-lock personal optical communicators;

- Optical terapop "phones";

- "Retina-code" communications;

- Computers using optical processing and possessing intelligence quotients (IQs) that exceed those of their individual human developers.

Retina-code communications (put forward as an idea for the first time here) could be particularly interesting because the use of human iris information as a security technique was already established by the end of the twentieth century. By the mid-1990s the United States National Security Agency (NSA) had developed this technique for security checks on personnel desiring to enter highly secure areas (e.g., the Menwith Hill base described in Chapter 4). Later this approach was proposed as an effective anti-crime technology for use with, for example, automatic cash machines and credit card checking.

With retina-code communications the approach would probably begin with automatically scanning the retinas of each person desiring to communicate. This information is then retained in a high-speed storage device and used to specify the sources and destinations of all messages.

This concept takes things several steps further on from retina-code security since the information is now being used as the ultimate determinant for personal communications.

The concept of the optical terapop (and optical terapop "phones") is described in Section 10.4.

10.3   Superhighway on-ramps of the twenty-first century

During the first decade of the twenty-first century technologies such as LMDS and the stratospheric systems (see Chapter 8), broadband satellite (see Chapter 9), and single-mode fiber optics all thrust forward—establishing their respective niches and growing to serve the ever-expanding needs of the information society. LMDS and MVDS networks are being implemented in many economies globally, several of the broadband satellite constellations will shortly be launched and will then come on-stream, and broadband single-mode fiber cables will increasingly be installed in the same pipes as the earlier CATV multimode cables of the 1980s and 1990s. The role of satellite systems will gradually become relegated to mainly nonmetropolitan areas and support for developing economies.

The numbers of broadband subscribers will grow rapidly, cumulatively, and globally, and as mankind enters the second decade of the twenty-first century optical technologies will begin to force the pace.

It is often considered that optical technologies preclude mobile applications, but this is actually far from true, as Table 10.1 shows.

The rationale underlying these features is as follows:

- The optical mobile applicability comes through the concept of the optical terapop (see Section 10.4).

Table 10.1
Wireless Technologies Compared

Parameter	Radio	Optical
Mobile application	Yes	Yes
Security	Secure	Highly secure
Reach distance	Medium-to-long-range	Short-range
Available bandwidths	Narrow-band or broadband	Extremely broadband
Maturity of the technology	Mature	Comparatively recent
Human safety	Some serious concerns	Relatively safe

- Optical signals are generally much harder to intercept than their electrical counterparts.

- The short range noted for optical wireless in Table 10.1 only applies to the free-space optical environment. This is not a problem because wherever long-range is required fiber is used.

- As noted in Chapters 2 and 3 there is an extremely wide bandwidth that extends to several tens of terahertz associated with the use of the optical spectrum.

- At present much of the available optical technology is relatively recent, but by 2020 this will cease to be the case. The technology will become very mature.

- Although as yet unproven and in all probability groundless, there do exist some serious concerns about the safety of using mobile phones during the transmission phase. Relatively speaking, optical systems are safe because lower power densities can be used, and in any event the electromagnetic radiation involved cannot penetrate human tissue.

The above comparisons indicate that optical wireless has many clear advantages over radio-based technologies. Although at the turn of the millennium optical wireless technologies are in their infancy, as far as full commercial exploitation is concerned, by the year 2020 they will almost certainly become the norm for wireless communications and fiber cable.

We now move on to consider what is being termed here the *optical terapop*.

10.4 The terabit junction, terapops, and bit-serial optical processing

In Chapter 3 the gigabit junction and the concept of GigaPOPs are described. In context these are distinctly applied to cabled networks operating up to gigabit rates.

The relatively futuristic concepts of the terabit junction and terapops are now introduced here and these are presumed to apply to highly broadband mobile optical networks. The basic concept of an "optical terapop"—that is, a point of presence (pop) within an optical network functioning at terabit rates—is indicated in Figure 10.3.

Practically everything in this scenario is operating on a highly broadband basis. The ultimate aim is to provide a secure, broadband, mobile cellular service using entirely optical technology. Any person within the group of people occupying the specific optical wireless communications cell shown in Figure 10.3 may make calls using an optical mobile phone. This connects with an optical transceiver that is in turn coupled through to the external broadband network (the work of Peter Cochrane of BT Labs in the United Kingdom is gratefully acknowledged for the basic concepts underlying Figure 10.3).

In practice the "optical mobile phone" would be a multimedia device with many functions, including Internet generation III surfing and communications.

Because it is potentially very significant for future optical communications systems, we now take a look at bit-serial optical processing (BSOP).

By the closing years of the twentieth century dense-wavelength division multiplexing (DWDM) was becoming increasingly well established in very high bit rate optical networks, and this is described in some detail in Chapter 3.

However, although it represents an advanced and extremely effective technology, DWDM remains an approach that cannot actually get anywhere near to processing the individual optical bits in the photonic bit

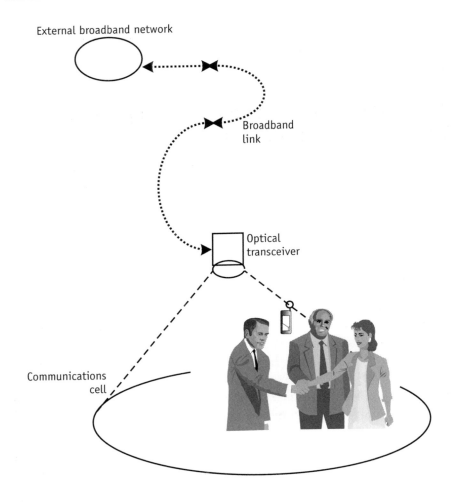

Figure 10.3   An optical terapop.

stream. At the DWDM level all one can do is to convert the signal back to an electrical version, process the information electronically, and then reconvert to optical. This process means that crosstalk and noise are added at each node, therefore reducing the signal-to-noise ratio and severely limiting the number of switching nodes that can be implemented.

Theoretically the processing could be optical, but this would require very large numbers of optical signal regenerators, that is, it represents an expensive solution.

Recent research has indicated the feasibility of all-optical switching devices capable of operating at switching rates up to 100 GHz and such devices switching at 40 GHz have been demonstrated by BT Labs in the United Kingdom. The acronym TOAD (terahertz optical asymmetric demultiplexer) has been coined for this class of devices by Paul Prucnal of Princeton University in the United States, who first demonstrated the TOAD.

In order to understand the basic principles of the TOAD, we first need to consider a more classical optical device called the Mach-Zehnder interferometer and a schematic view of this is indicated in Figure 10.4(a). The input light wave signal is fed into a 50:50 optical power splitter, providing two identical light streams within the device. These optical streams are then fed into another splitter that simply passes the two signals, now A and B, out of the device.

The most important function of this device is to provide two light outputs with differing phases, with the entire device fed by one light signal input. The different phases, shown by the two time-shifted sinusoids for outputs A and B in Figure 10.4(a), occur as a result of the naturally different time delays for the two separated signals traveling along upper and lower routes. The device can be biased so that all of the light exits from one or other of the two output ports and indeed the signal can be switched between output ports.

It is well known in optical research circles that such Mach-Zehnder devices can be switched at tens of gigahertz rates. The technology uses entirely commercially available devices and the optical paths may be fibers or integrated optical waveguides formed on a chip.

A TOAD optical switch is a variation on the Mach-Zehnder principle and the general arrangement is shown in Figure 10.4(b). In the TOAD device the key element is the semiconductor optical amplifier (or SOA) that functions as an optical switch.

The main 50:50 power splitter [to the left hand side of Figure 10.4(a)] has all four ports in operation and the input signal is fed into the top port. At the output port on the opposite side of this splitter the optical signal is

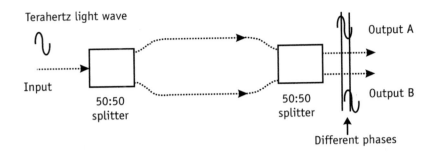

(a) The Mach-Zehnder interferometer

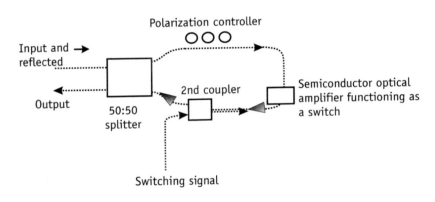

(b) A TOAD optical switch

Figure 10.4 A high-speed TOAD (terahertz optical asymmetric demultiplexer).

passed along a polarization-controlled fiber or waveguide and into the SOA. The SOA output feeds the signal along fiber or waveguide again and into a second coupler. The only function of this second coupler is to pass this signal on toward the main coupler at the same time as enabling a switching signal to pass in the opposite direction in order to control the SOA.

It is important to offset the position of the SOA in the transmission loop so that clockwise and anti-clockwise pulses reach the device at precisely specified differing times and this effect determines the switching characteristic. The output from the main coupler is then finally the switched version of the original optical signal and switching rates can be at least as high as 100 GHz.

There is also some extraneous reflected power from the input port but this is small in comparison with the output and can readily be absorbed.

Basic all-optical TOAD memories have been demonstrated by BT Labs and there seems to be little doubt that system-level bit-serial optical processors (BSOPs) will emerge as commercially available devices based upon this technology.

10.5   The brilliant broadband future

Technologies like those described above (notably BSOPs) and throughout this book are leading to profound changes in the way we all work and play. The increasingly intense interaction between computer-like technologies and communications had, well before the close of the second millennium, led to the functional concept of virtuality. During the 1980s a highly commercial and widespread interest in virtual reality (VR) grew steadily, having its origins in the simulation technologies that are so well known in their aircraft and car-driver learning-mode manifestations. Videoconferencing inevitably followed teleconferencing as a means for minimizing physical travel at the same time as flexibly bringing together several people at any time—and from almost any location globally. There were, however, some serious practical limitations that inhibited the commercial exploitation of these otherwise revolutionary approaches:

- Relatively poor-quality images with fuzzy motion rendering;

- Doubts about the inherent security of the systems;

- Lack of appreciation concerning time-zone restrictions in different parts of the world.

The implementation of broadband, increasingly optical, networks will practically eliminate the first two problems. As far as time-zone differences are concerned, experience will greatly assist as more and more people become used to communicating internationally. Another factor that will help in this regard could well be the use of digital memories with extremely fast real-time processing and high-speed access. BSOPs may well prove to be the enabling technology here.

Eventually, physical travel may become almost entirely relegated to history with only the occasional microlight and other solar-powered craft taking people on vacations. The World Wide Web never sleeps. It is a 168-hr per week, every week, absolutely continuously available network. As suggested in Chapter 7, eventually everyone's home may become essentially a teleport.

For corporations the historical trends are illustrated in Figure 10.5, covering the period from the 1960s well into the twenty-first century. At first companies were all essentially "stand-alones," that is, each firm had its headquarters offices and operational facilities all on the same site or at least within a fairly localized zone. By the 1970s corporations were segmenting their increasingly varied activities into different specialized divisions, although maintaining coordination between these divisions (ensuring a corporate identity) became a serious challenge. During the 1980s this challenge was largely met by having an integrating "umbrella" holding company that interacted efficiently with the highly specialized divisions, ensuring that a cohesive operation was driven forward.

By the final decade of the twentieth century the situation had become much more complex and it was frequently difficult for an outsider (even an insider!) to unravel and understand the precise structure and interrelationships within many corporations. This was and still is in many instances the era of cooperating corporations, with operations such as joint ventures, consortia, and many other joint activities being pervasive.

All successful companies appreciate that efficient and secure communications hold the key to continued prosperity, and e-commerce now represents the latest manifestation of this knowledge. Since the middle of the twentieth century this fact has become increasingly evident to most corporations of all sizes, but even at the end of the century almost all communications were narrowband—certainly by the standards of the coming broadband satellite, terrestrial, and optical technologies. As described elsewhere in this book the limitations of these narrowband links severely restricts intercompany communications and also external trade.

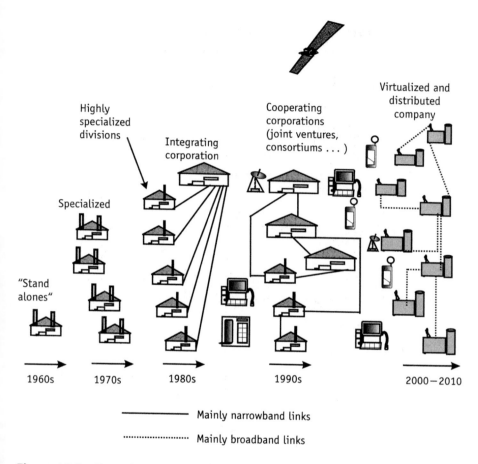

Figure 10.5   Toward the virtualized digital company.

As we enter the twenty-first century—the third millennium—so many of the broadband systems described will be coming increasingly on-stream. By 2020 this will mean the continued evolution of virtualized and distributed companies and, as with many late twentieth century organizations, these will also be global operations.

This type of company is definitely a virtual one—distributed globally and having elements ranging from one-man-bands to massive previrtual corporate divisions with totally automated manufacturing, warehousing, and servicing facilities. Most will probably be completely service-oriented organizations that maximize the implementation of efficient software to provide extremely well matched services to each customer. Individual customers, located anywhere in the world, will be treated as if they were the only customers of the company, and trade will therefore be greatly enhanced.

The concept of the global virtual company is further illustrated in Figure 10.6 where entirely broadband interconnections are indicated. The work of Peter Cochrane (BT Labs, United Kingdom) is gratefully acknowledged for the basic concepts underlying both Figure 10.5 and Figure 10.6.

Users will activate many kinds of "gizmos" to match their needs at any time; examples are taken directly from the list provided earlier in this chapter:

- Personal electronic and photonic assistants (PEPAs);
- Auto-scan-and-lock personal optical communicators;
- Optical terapop "phones";
- "Retina-code" communications;
- Computers using optical processing and possessing intelligence quotients (IQs) that exceed those of their individual human developers.

By the year 2020 the world may well appear chaotic and highly self-organized, with practically all knowledge being acquired via technological means—principally Internet III.

The scenario is truly one in which broadband communications, functioning at gigabits per second and terabits per second rates, provide an indispensable backbone network that links directly to the users.

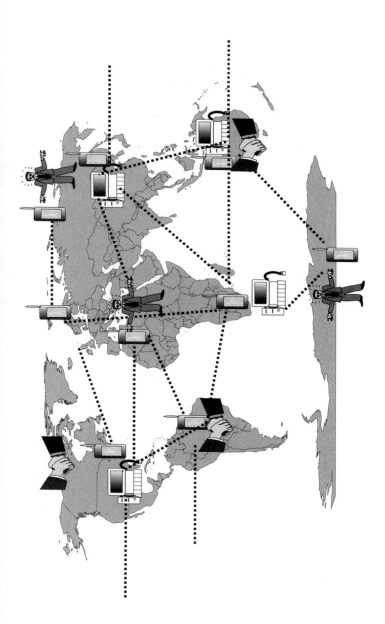

Broadband interconnections using fiber, satellite, terrestrial, and other wireless.

Figure 10.6 The global virtual company.

Select bibliography

Cameron, I., "E-commerce: Beyond the Physical World," *Electronics Times*, June, 1999, p. 26.

Cochrane, P., "Commerce 2016," in: *The Electronics Industry Debate*, The Institute of Directors, Pall Mall, London, October 1996 (unpublished).

Dettmer, R., "The 100 GHz Light Switch," *IEE Review*, March 1999, pp. 69–71.

Straunik, A., "Gadgets Galore," *Information Week*, June 2, 1999, pp. 33–34 (http://www.informationweek.co.uk).

Walko, J., "Forward Planner," *Electronics Times*, November 2, 1998, p. 32.

List of Acronyms and Abbreviations

ACC analog control channel

ADSL asymmetric digital subscriber line

AGC automatic gain control

AM amplitude modulation

AN aligning units

ARIC ATM radio interface card

ART Advanced Radio Telecom

ATM asynchronous transfer mode

ATSC Advanced Television Systems Committee

AUG   administrative unit group

AWACs   Airborne Early Warning AirCraft Systems

B-CDMA   broadband code-division multiple access (cf: W-CDMA)

BER   bit error rate

BFN   beam forming network

B-ISDN   Broadband ISDN

BPSK   binary phase-shift keying

BSOP   bit-serial optical processors

BT   British Telecom

Calren2   California Research and Education Network 2

CATV   cable television

CIS   Commonwealth of Independent States

COC   chain of command

COFDM   coded orthogonal frequency division multiplexing

COMINT   Communications Intelligence

COTS   commercial-off-the-shelf

CPE   customer premises equipment

CRT   cathode-ray tube

CSMA/CD   carrier-sense multiple access with collision detection

CW   continuous wave

DARPA   Defense Advanced Research Projects Agency

DASA   Daimler Chrysler Aerospace

DBS   direct broadcast by satellite

DECT Digital European Cordless Telecommunications

DFB distributed feedback

DoD Department of Defense

DRAM dynamic random access memory

DS digital signal (or direct sequence)

DSP digital signal processing

DSCS Defense Satellite Communications System

DTH direct-to-home

DTV digital television

DVB Digital Video Broadcasting

DWDM dense-wavelength division multiplexing

EDFA erbium-doped fiber amplifier

EHF extremely high frequency

ELINT electronic intelligence gathering

EPC electronic power conditioner

ETSI European Telecommunications Standards Institute

EW electronic warfare

FCC Federal Communications Commission

FDDI fiber distributed data interface

FDM frequency-division multiplexing

FDMA frequency division multiple access

FDMA/SCPC frequency division multiple access/single channel per carrier

FM   frequency modulation

FT   *Financial Times*

FTTH   fiber-to-the-home

FTTO   fiber-to-the-office

Gbps   gigabits per second

GEO   geosynchronous earth orbit

GESN   Global EHF Satellite Network

GHz   gigahertz

GigaPOP   giga points of presence

GII   global information infrastructure

GMF   ground mobile forces

GNSS   Global Navigation Satellite System (U.S.S.R., now C.I.S.)

GPS   Global Positioning Satellites

GSM   global system for mobile telecommunications

HALO   High Altitude Long Operation

HDTV   high-definition television

HEO   highly elliptical orbiting

HF   high frequency

H-F/C   hybrid fiber/coax

HNS   Hughes Network Systems, Inc.

HQ   headquarters

HSCI   Hughes Space and Communications International, Inc.

HTML   Hypertext Markup Language

IC   integrated circuit

ICO   INMARSAT Council; ICO Communications Ltd.

IDR   intermediate data rate

iDTV   interactive digital television

IEEE   Institute of Electrical and Electronics Engineers

IMT   international mobile telecom

INFOSEC   information security

IP   Internet Protocol

ISDN   Integrated Services Digital Network

ISI   intersymbol interference

ISL   intersatellite links

ISM   industrial, scientific, and medical

ISP   Internet service provider

ITU   International Telecommunications Union

IXC   international exchange

JDW   *Janes Defense Weekly*

KC   Kingston Communications

kHz   kilohertz

KMI   Kessler Marketing Intelligence (a U.S. analyst company)

KSS   Kingston Satellite Services

LAN   local area network

LD   long distance

LEC   local exchange carrier

LEO   low earth orbit

LLC   logical link control

LMDS   local multipoint distribution services

LNA   low-noise amplifier

LNB   low-noise blocks

MAN   metropolitan area network

MBA   main beam array

Mbps   megabits per second

MCM   multichip modules

MDU   multidwelling unit

MEO   medium-earth orbiting satellite

MHz   megahertz

MILSATCOM   Military Satellite Communications

MMDS   multipoint microwave distribution system

MMIC   monolithic microwave integrated circuits

MMT   Marconi Materials Technology

MMW   millimeter-wave

MOB   main operating base (NATO)

MOST   military off-the-shelf satellite terminal

MPIM   monolithic photonic integrated modules

MREN   Metropolitan Research and Education Network

MSC   Multimedia Supercorridor

MSS   mobile satellite system

MUX multiplexing

MVDS microwave video distribution services

MW microwave

MWMIC millimeter-wave microwave integrated circuits

MWS multimedia wireless system

NCNI North Carolina Networking Initiative

NICS NATO Integrated Communications System

NIVR Netherlands Agency for Aerospace Programs

NRZ non-return to zero

NSA National Security Agency

NTT Nippon Telegraph and Telephone Corporation

N-VOD near-video on demand

OADM optical add/drop multiplexers

OLS optical line system

OMP Osaka Media Port

OMT Osaka Media Teleport

PC personal computer

PCM pulse code modulation

PCS personal communications service

PEPA personal electronic and photonic assistants

PHS personal handyphone system (Japan)

POTS plain old telephone service

PSK phase-shift keying

PSTN   public switched telephone network

PSU   power supply unit

PVC   private virtual circuit

QAM   quadrature amplitude modulation

QML   Qualified Manufacturers List

QPL   Qualified Parts List

QPSK   quadra-phase shift keying

RAM   random access memory

REN   Research and Education Network

RF   radio frequency

ROI   return on investment

SAW   surface acoustic wave

SBIR   space-based infrared systems

SCPC   single channel per carrier

SD   short-distance

SDH   synchronous digital hierarchy

SDTV   satellite distributed TV; standard definition TV

SHF   super high frequency (3–30 GHz)

SIGINT   signals intelligence

SiGe   silicon-germanium

SM   surface mount

SMATV   satellite master antenna television

SMDS   switched multimegabit digital service

SMR   specialized mobile radio

SOA   semiconductor optical amplifier

SOIC   small outline IC

SONET   synchronous optical network

SSPA   solid-state power amplifiers

STM   synchronous transport module

Tbps   terabits per second

TC8PSK   trellis-coded quarternary phase-shift-keying

TCM   Thomson-CSF Microélectronique

TDM   time-division multiplexing

THz   terahertz

TOAD   terahertz optical asymmetric demultiplexer

TPN   *Telecom Product News*

TSSR   tropospheric and satellite services radio

TWT   traveling wave tube

UAE   United Arab Emirates

UHF   ultra high frequency (300 MHz–3 GHz)

UMS   United Monolithic Semiconductors

UMTS   Universal Mobile Telecommunications Systems

USSB   U.S. Satellite Broadcasting

VHF   very high frequency (30 MHz–300 MHz)

VLA   Very Large Array

VLAN   virtual local area network

VPN   virtual private network

VSAT   very small aperture terminal

VSB   vestigial sideband

WAN   wide area network

WAP   Wireless Access Protocol

WARC   World Administrative Radio Conference

WCN   World Computer Network

W-CDMA   wideband code-division multiple access (cf: B-CDMA)

WDM   wavelength division multiplexer

WHCA   White House Communications Agency

WLAN   wireless local area networks

WLL   wireless local loop

WRC   World Radio Conference

WTA   World Teleport Association

xDSL   digital subscriber line (the x signifies *any* method)

XVGA   extra video graphics adaptor ($>1,024 \times 768$ pixels)

Y2K   year 2000

About the Author

Terence C. Edwards is the executive director of Engalco, a consultancy firm based in the United Kingdom mainly specializing in signal transmission technologies and the global industry. He holds an M.Phil. postgraduate degree in microwave research, has led seminars on fiber optics, and has written several articles and books including: *Fiber Optic Systems—Network Applications* (John Wiley and Sons, 1989), *Foundations for Microstrip Circuit Design* (John Wiley and Sons, 1st edition, 1981; 2nd edition, 1991), and *Microwave Electronics* (Edward Arnold, 1984). He has been involved with and managed many consultancy programs. Studies completed and published by Engalco include several on the microwave industry in both Europe and North America and, under sponsorship with a North American corporation, microwave-based systems in Asia, Africa, and Latin America. Terry travels considerably to conferences and trade exhibitions including Atlanta, Denver, Geneva, Paris, Philadelphia, San Jose, and Santa Rosa in recent years. He is on the editorial advisory board for the *International Journal of Communication Systems*.

Terry Edwards has also acted as an expert witness at a Philadelphia court hearing involving a fiber-optics product technology dispute. A Fellow of the IEE and member of the IEEE, Terry regularly consults for both national and overseas companies and is on the prestigious IEE (London) President's List of Consultants.

He may be reached via electronic mail at terryengalco@compuserve.com.

Index

Recent Titles in the Artech House
Space Technology and Applications Series

Bruce R. Elbert, Series Editor

Gigahertz and Terahertz Technologies for Broadband Communications, Terry Edwards

Introduction to Satellite Communication, Second Edition, Bruce R. Elbert

Low Earth Orbital Satellites for Personal Communication Networks, Abbas Jamalipour

Mobile Satellite Communications, Shingo Ohmori

The Satellite Communication Applications Handbook, Bruce R. Elbert

Understanding GPS, Elliott D. Kaplan, editor

For further information on these and other Artech House titles, including previously considered out-of-print books now available through our In-Print-Forever® (IPF®) program, contact:

Artech House
685 Canton Street
Norwood, MA 02062
Phone: 781-769-9750
Fax: 781-769-6334
e-mail: artech@artechhouse.com

Artech House
46 Gillingham Street
London SW1V 1AH UK
Phone: +44 (0)171-973-8077
Fax: +44 (0)171-630-0166
e-mail: artech-uk@artechhouse.com

Find us on the World Wide Web at:
www.artechhouse.com